Space Policy Alternatives

Published in cooperation with
the Center for Space and Geosciences Policy,
University of Colorado, Boulder

Space Policy Alternatives

EDITED BY
Radford Byerly, Jr.

Westview Press
BOULDER • SAN FRANCISCO • OXFORD

Westview Special Studies in Science, Technology, and Public Policy

This Westview softcover edition is printed on acid-free paper and bound in library-quality, coated covers that carry the highest rating of the National Association of State Textbook Administrators, in consultation with the Association of American Publishers and the Book Manufacturers' Institute.

All rights reserved. No part of this publication may be reproduced or transmitted in any form or by any means, electronic or mechanical, including photocopy, recording, or any information storage and retrieval system, without permission in writing from the publisher.

Copyright © 1992 by Westview Press, Inc.

Published in 1992 in the United States of America by Westview Press, Inc., 5500 Central Avenue, Boulder, Colorado 80301-2877, and in the United Kingdom by Westview Press, 36 Lonsdale Road, Summertown, Oxford OX2 7EW

Library of Congress Cataloging-in-Publication Data
Space policy alternatives / edited by Radford Byerly, Jr.
    p. cm. — (Westview special studies in science, technology, and public policy)
  Includes bibliographical references and index.
  ISBN 0-8133-8618-7
  1. Astronautics and state—United States. I. Byerly, Radford.
II. Series.
TL789.8.U5S587 1992
338.9′4′0973—dc20   92-25347
                                                                                         CIP

Printed and bound in the United States of America

∞ The paper used in this publication meets the requirements of the American National Standard for Permanence of Paper for Printed Library Materials Z39.48-1984.

CONTENTS

Preface vii
List of Contributors ix

1   Introduction, Radford Byerly, Jr. 1

PART ONE: OVERVIEW 11

2   The Future of the Unmanned Space Program
       Albert D. Wheelon 13

3   The Lonely Race to Mars: The Future of Manned
       Spaceflight
       Alex Roland 35

PART TWO: CONTEXT 51

4   The NASA Budget: For Whom, For What, and How Big?
       Molly K. Macauley 53

5   NASA and the Budget Process
       Michael L. Telson 77

6   Decision Making and Accountability in the U.S. Space
       Program: A Perspective
       Jack D. Fellows 93

7   The Role of Incentives and Accountability in Industry
       and Government
       Angelo Guastaferro 107

8   Policy Issues Pertaining to the Space Exploration
       Initiative
       Maxime A. Faget 117

9   Low Cost Access to Space for Small Science and
    Technology
    Paul J. Coleman, Jr. 125

PART THREE: PROGRAMS 137

10 NASA's Space Science Program: The Vision and the
    Reality
    Richard McCray and S. Alan Stern 139

11 And Then There Was One: The Changing Character
    of NASA's Space Science Flight Program
    S. Alan Stern and M. Jay Habegger 167

12 EOS, the Earth Observing System: NASA's Global
    Change Research Mission
    Ferris Webster 183

13 The Future of the Space Station Program
    Ronald D. Brunner, Radford Byerly, Jr., and
    Roger A. Pielke, Jr. 199

14 The Space Shuttle Program: "Performance Versus
    Promise"
    Roger A. Pielke, Jr., and Radford Byerly, Jr. 223

PART FOUR: CONCLUSIONS 247

15 Imagined Frontiers: Westward Expansion and the
    Future of the Space Program
    Patricia Nelson Limerick 249

16 Can the United States Conduct a Vigorous Civilian
    Space Program?
    Radford Byerly, Jr. 263

Index 275

PREFACE

In 1989 I edited a somewhat similar group of essays published by Westview under the title *Space Policy Reconsidered*. The preface to that volume began with this statement:

> For some time space policy debate has been too constrained by preexisting assumptions and programs. There is also a related need for a community of independent space policy analysts in order to inform those discussions. The aim of this book is to take a step toward meeting such needs.

That statement is repeated here because it is still valid — and this book is intended to address the same unmet needs.

Initial drafts of these contributions were prepared before the Augustine Commission released its excellent report and, although they were revised afterwards, for the most part they do not reflect that report. Part of the reason for this is that although Augustine could and perhaps should have initiated a thoughtful debate on space policy, it did not. As discussed in the Introduction, the Augustine report has had much less impact than one would have thought, although it is often referenced. Thus this volume is submitted as relevant material for debate yet to come.

The previous volume was somewhat introspective; it focused on space activities and described programs and problems. In contrast, the present volume attempts two new departures: to look more to the future and to look outside space programs to the policy context in which they operate. It attempts, where possible, to prescribe solutions, to offer alternatives for consideration.

Where the book is successful, credit belongs to the chapter authors; where it is not, responsibility falls on me. In addition to the authors, others too numerous to mention have contributed. However, I must gratefully acknowledge and thank Patricia Duensing, who produced the text. I thank all my collaborators and helpers.

Finally, the support of the Alfred P. Sloan Foundation and of the University of Colorado is gratefully acknowledged.

Radford Byerly, Jr.

CONTRIBUTORS

Ronald D. Brunner is a Professor of Political Science at the University of Colorado and has been Director of the Center for Public Policy Research. Currently studying the Space Station, he has previously worked on energy and social welfare.

Radford Byerly, Jr., is currently Chief of Staff of the Committee on Science, Space, and Technology of the House of Representatives. When this book was started he was Director of the Center for Space and Geosciences Policy at the University of Colorado.

Paul J. Coleman, Jr., is a Professor of Space Physics and Director of the Institute of Geophysics and Planetary Physics at the University of California at Los Angeles. He is also President of the Universities Space Research Association.

Maxime A. Faget was employed by NASA for more than 35 years. As Director of Engineering and Development at NASA's Johnson Space Center, he was instrumental in the design and development of all U.S. manned spacecraft programs from Mercury to the Space Shuttle. He currently serves as Chairman of the Board for Space Industries International, Inc.

Jack D. Fellows is currently a Senior Program Analyst for space and science issues in the Executive Office of Management and Budget. Prior to the Executive Office, he was a Research Associate at the University of Maryland's Remote Sensing Laboratory and an American Geophysical Union Science Fellow in the U.S. Congress.

Angelo Guastaferro is currently Vice President of NASA Programs at Lockheed Missiles & Space Company. Prior to this he had a distinguished career at NASA, particularly as an executive in the planetary program and as Deputy Director of the Ames Research Center.

M. Jay Habegger is a graduate student in telecommunications at the University of Colorado. His research interests include science and technology policy and telecommunications regulation.

Patricia Nelson Limerick is a Professor of History at the University of Colorado. Her historical research has focused on the

American West, resulting in two books, *Desert Passages* and *Legacy of Conquest*, and a collection, *Trails: Toward a New Western History*.

Molly K. Macauley is an economist at Resources for the Future, where she directs a research program on space economics. She has written on the economics of space transportation and of the geostationary orbit.

Richard McCray is Professor of Astrophysics at the University of Colorado at Boulder. His research areas are theoretical astrophysics and space astronomy, and he has served on several scientific committees concerned with space science programs and policies.

Roger A. Pielke, Jr., is a doctoral student in Political Science at the University of Colorado. His research interests include science and technology policy, symbolic politics, and American politics.

Alex Roland is Professor of History at Duke University, where he teaches military history and the history of technology. From 1973 to 1981 he was a historian with the National Aeronautics and Space Administration.

S. Alan Stern is an astrophysicist and planetary scientist with experience in industry, academe, and NASA. Previously a member of the Center for Space and Geosciences Policy at the University of Colorado, Dr. Stern is now a Principal Scientist in the Space Science Department at Southwest Research Institute in San Antonio.

Michael L. Telson presently serves as Senior Budget Analyst with the Committee on the Budget of the U.S. House of Representatives, responsible for Federal energy, science, and space programs. Since 1975 he has covered energy policy and other areas for the Committee. His Ph.D. is in electrical engineering.

Ferris Webster is Professor and Director of the Oceanography Program in the College of Marine Studies at the University of Delaware. He is a physical oceanographer, involved in studies of the effect of the ocean on global climate.

Albert D. Wheelon has had a distinguished career in government and industry which has spanned space activities from intelligence satellites to commercial communication satellites. Recently he retired as Chairman of the Board of Hughes Aircraft Corporation.

Chapter 1
INTRODUCTION

Radford Byerly, Jr.

WE ARE HANDICAPPED BY POLICIES BASED ON OLD MYTHS RATHER THAN CURRENT REALITIES.
—J. W. FULBRIGHT

The strongest myths handicapping the space program have to do with the Apollo program. The simplest is that Apollo was the optimal way to run a space program — which is based on selective memory of the budget ramp-up in the early sixties and suppression of the drastic budget cuts in the late sixties, cuts that began even before the Moon landings.

Another myth, more subtle and therefore more pernicious, is that Apollo was a space program. It was a space mission, but it was a geopolitical program: Its origins and purpose were geopolitical. Apollo's purpose, its goal or *end*, was to show the rest of the world — many in the US and USSR already knew — that the US was *the* leader in high technology, particularly rocketry and the related ability to deliver atomic weapons.

Going to the Moon was the *means*.

Since Apollo it seems that the space program has never really gotten its footing here on Earth. Apollo was clearly a success both geopolitically and as a space mission. Yet the agency was "punished" by having its budget slashed. There may be an unconscious myth that even magnificent success will not be rewarded.

At any rate, times have changed and the old way of doing business isn't going to work. This has been thoroughly discussed elsewhere,[1] so now it is sufficient to note that large deficits have reduced the chances for budgets to grow and the collapse of the USSR has taken away what has been our biggest motivator. And the program is floundering: It is overdependent on the fragile shuttle; Station stutters and splutters; missions overrun budgets and then underachieve. It is a program badly in need of a fresh outlook, some new myths to replace the existing ones, a new way of doing business. (Not all of the myths hampering the space program are as old as Apollo. For example, the Shuttle is presented to the world as a symbol of advanced technology and engineering, yet as Pielke and Byerly show in their chapter, and as most insiders recognize, it has never met its original goal of low-cost, highly-operational access to space. Even now new false myths are being created around the Space Station, as Brunner et al show. There is need to move rapidly.)

A New Way of Doing Business

This new style is struggling to be born. Two hundred years from now historians will say that a fundamental change in space policy occurred after the Challenger accident. But we are in the middle of the change and can see no pattern: there is flux but much is still unmoved. Reversals occur. The future is not clear. But poor program performance shows both that the methods of the past are inadequate, and also that they are not yet dead. Collectively we need to define and install a new paradigm for the space program.

We have called the old way of doing business "The Apollo paradigm."[2] Roland and others have described the new era as beginning with the 1986 Challenger accident,[3] so we can call the as-yet-undefined new way of operating "the post-Challenger paradigm."

The Apollo paradigm is primarily characterized by a focus on transportation, i.e. getting into space and thence on to the Moon.[4] Of course simply getting into orbit was the first problem we had to overcome. And a hard problem it is — the kinetic energy of a mass in orbit is larger than the energy of that same amount of high explosive: In order to put a fragile spacecraft into orbit one has to carefully accelerate it until it has the energy of a bomb. The old paradigm has other typical characteristics:

— It attracts people driven by a higher goal, a *vision* of space exploration as man's destiny.
— It focusses on manned spaceflight.

— It gravitates toward missions that are large in scope, scale, cost, and duration. The missions are also composed of coupled, interdependent elements; centralized; highly planned and engineered.
— Further, these highly interdependent programs are often planned with little margin to accommodate unexpected changes, e.g., budget shortfalls or accidents: This is called "success orientation" and leads to programmatic fragility.
— Cost is not treated as an economic consideration, but as a political one; that is, the operative question is "Will Congress appropriate the money?" not "What else could be done with the money?" That is, there is no consideration of opportunity cost.
— Often there is no clear *practical* original justification for missions. Rather the mission usually emerges first, either as part of the vision or because new technology makes it possible; subsequently practical applications are sought, especially as justifications. The mission itself — the hardware and its launch — are paramount. Thus arise inversions of ends and means.

Many of these characteristics originated in Apollo's geopolitical purpose. For example, no cost was too great to beat the Soviets. For example, many in the space program have never realized that going to the Moon was a means, not an end, and they continue to mix the two.

The Shuttle, Space Station Freedom, and the "90-day study," referred to as the Human Exploration Initiative,[5] are examples of the old mode of thinking and operating. Luckily the 90-day plan was recognized to be impractical and was rejected.[6]

The post-Challenger paradigm remains to be conclusively articulated. That may be done only in hindsight, but several likely characteristics are emerging in thoughtful discussions of the future of the space program. The Augustine commission made several recommendations that would change the character of the U.S. civil space program.[7] For example, some of their recommendations are:

- That the civil space science program should have first priority for NASA resources. . . [p. 25]
- That the multidecade set of projects know as Mission to Planet Earth be conducted as a continually evolving program rather than as a mission whose design is frozen in time. . . [p. 27]
- That the Mission from Planet Earth be configured to an open-ended schedule, tailored to match the availability of funds. [p. 28]

- That NASA . . . reconfigure and reschedule the Space Station Freedom [for only two missions, life sciences and microgravity]. In so doing, steps should be taken to reduce the station's size and complexity . . ." [p.29]

Let us consider each of these. If NASA were to make science top priority that would be a major reversal from the emphasis on manned spaceflight and on missions for their own justification. Dutton *et al* have made a broader recommendation, i.e. that the pursuit of information,[8] which encompasses not only scientific but also technical research and could even include some human exploration, should be a goal co-equal with fostering national prestige through manned programs. The point is, adoption of top priority for science would end the dominance of transportation and hardware.

The recommendations regarding Mission to Planet Earth (MtPE), Mission from Planet Earth (MfPE), and Station amount to a repudiation of the philosophy of the large-scale, tightly integrated, centrally planned, cost-is-no-object approach to space projects. The recommendation for a smaller-scale approach to MtPE is echoed by the Frieman engineering study of the proposed Earth Observing System[9] and by the August 1991 report of NASA's Space Science and applications Advisory Committee,[10] both of which saw the value of smaller missions. A similar new philosophy is advanced by several authors in this volume, Coleman, McCray and Stern, Webster, and Brunner et al in particular.

One of the themes running between the lines through the Augustine report is that the space program needs to be more flexible, robust, and resilient: It is programmatically fragile. For example it is "over committed in terms of program obligations relative to resources available" and "overly dependent on the Space Shuttle for access to space".[11]

Thus the recommendation that MfPE operate on a go-as-you-pay basis is an attempt to reduce financial overcommitment and to recognize that cost must be a factor. To implement such a recommendation would also force the design of decentralized programs composed of relatively independent modular elements in contrast to the approach characteristic of the Apollo paradigm.

Overdependence on the Shuttle was addressed earlier by the two Welch reports.[12] The first, issued shortly after the Challenger accident recognized the Shuttle as a unique and valuable resource which should be conserved. This report recommended reversing the present philosophy of flying the Shuttle as often as possible and instead conserving it for missions that need its unique capabilities. The second Welch report recommended a "robust space transportation capability" to be achieved by planning and acquiring adequate resources "to absorb unexpected but inevitable delays." This is an example of an attempt to move away from the success-orientation of the Apollo paradigm.

The point of referring to these reports and recommendations is that a new way of doing business is known, is public, and is being recommended to the leaders of the space program. It is, however, also being resisted.[13] For example although the Welch reports were clearly aimed at preserving the Shuttle as a national resource, they are viewed by many as anti-shuttle, ie as an attempt to deflate the myth. NASA's present policies nominally recognize the Welch recommendations, yet e.g. for new payload flight assignments the burden of proof lies with those who do *not* want to fly on Shuttle.

Further examples of resistance to a new order are: The desperate grip on the impractical configuration for Space Station; the lack of concern over costs – funding is still the independent variable; Shuttle launches are still treated as "free";[14] there was strong internal preference to fly the large version of EOS despite the Freeman report and other indications that smaller is better. In fact the struggle to free EOS from giantism, described in this volume by Webster, epitomizes the space program's struggle to escape the Apollo paradigm.

It may be necessary for the White House or the Congress to force the space program to change its approach. If so, the mechanism may be an insistence that the program deliver what it promises when it "sells" its projects. Such a new departure – performance as promised – has been described by Brunner[15] and could become the new paradigm for the space program.

The Continuing Non-debate

The report of the Augustine committee, being comprehensive, thoughtful, progressive, directed to the White House, – and, not the least, full of common sense – presented a wonderful opportunity for debate on a new way of doing business; but no one seized it. Just as no one really debated many of the questions raised by the Challenger accident. That accident was disturbing enough to allow changes but there were no alternatives on the table at the time. The only significant change as result of the accident is that we now launch a few spacecraft on ELVs, i.e. the "mixed fleet." But we still fly the Shuttle as often as possible, sometimes launching satellites that could go on ELVs (such as TDRS), and the program is still highly dependent on Shuttle (e.g. Station is totally dependent).

When Augustine put his alternatives on the table there was no tragedy to unsettle the system, although there were a great many disturbing problems. We could have debated the go-as-you-pay principle for Mission from Planet Earth, we could have debated the need for a new launch system, and, as mentioned above, we could have debated whether science should be given top priority. It seems likely that such an approach would have resulted in an endorsement of and guidance on how to approach a mission to Mars, the next big step in space.

This is reminiscent of an earlier incident in the Station program when there was an attempt by our foreign partners to have their participation in Station recognized and guided by a formal treaty. A treaty would have required Senate ratification, which might have stimulated informed discussion of the Station and clarification of Congressional commitment to it. But debate was avoided.

If open, informed, and honest debate on policy can not only strengthen commitments but also be a source of new ideas.

What will it take to permit and stimulate debate? The answer is not evident. The June, 1991 debate over Space Station cancellation did not work: First, it was nominally about termination of the Station program although more fundamentally about support for manned spaceflight. Thus it was skewed, oblique, off-target.[16] Second, it was not well-informed, in part because of lack of time, in part because there were no constructive alternatives on the table. Further, the June vote was to some extent a power struggle within the House of Representatives between the authorizing committee and the appropriating committee, in which argument about the future of our space program was a means, not an end.

Will there be another attempt to terminate the Space Station program in 1992? Probably. Perhaps if the Station is terminated there will be an opportunity for constructive debate on the future of the space program. For this to occur we need to avoid a negative, emotional hangover in the form of a "stab-in-the-back" conspiracy hypothesis. Or another Shuttle accident could stimulate debate.

However if there is loss of life the context is again likely to be highly emotional and not conducive to clear thinking.

Perhaps the Augustine report did not stimulate debate because it focussed largely on programs and gave relatively less attention to the policy context. As an example of the policy context consider that the National Space Council (NSpC), chaired by Vice President Quayle, is locked in a power struggle with NASA — a struggle having more to do with who is going to make civil space policy than with the content of the policy. The two groups continue to wage their struggle within the confines of the Administration (perhaps because it is more about process than content and therefore too embarrassing to be revealed outside the beltway). Another example is the deficit-driven budget process discussed by Telson.

In sum, prospects for open, constructive discussion of space policy seem remote.

However, given the multitude of interests and the absence of philosopher kings such discussion seems necessary to achieve good policy. Thus there is a need for more than mere tolerance of debate; it must be encouraged. For example there is a need to exhume and examine the old, unconscious assumptions and to discard the obsolete ones. Informed debate can improve the content of policy — if NASA would create and share information on options — and the process can legitimize the outcome.

Key to an informed debate is open discussion of options and alternatives. For space policy this is hampered by a peculiar decision-making process. That is, in the civilian space program decisions are made (from middle levels upwards) by an advocacy system. Rather than developing and fairly weighing an array of different valid alternatives, the accepted, typical procedure is to select one approach and then to "sell" it to the next level of management. The choice may be rejected or accepted and, perhaps with slight changes, sold to the next level. Often even top NASA managers are not presented with alternatives, but, instead, with a binary choice to accept or reject one approach. This internal shortcoming is compounded by the fact that, unlike other policy areas such as energy or environment, there are no significant, independent, technically credible external sources of policy alternatives. Thus the option presented by NASA to the White House and to Congress must be accepted or rejected. This does not stimulate constructive debate about the best way to go, bur rather debate over whether or not to "kill" a program element.

If and when the country decides to grapple fundamentally with the future of its space program this book may make a contribution. The context for our space program is changing dramatically. The Federal budget is running a large deficit which so far has proven uncontrollable. At the same time potential solutions to domestic problems remain unfunded – a situation reminiscent of the years after Apollo when the program's budget plummeted. Meanwhile our old foe and space rival, the USSR, may literally cease to exist and our allies, Japan and Western Europe, are capable of becoming our equals in space. And a new policy player, the National Space Council struggles for a policy-making role. Change is in the air, if still incomplete. Unexpected opportunities may arise, and we should be prepared.

Organization of the Book

The book is constructed in four parts: In all parts each author was encouraged to look ahead, to recommend what should be done differently, in essence to suggest characteristics of the new paradigm for our space program.

Part I consists of two overview chapters – one covering manned programs (Roland), the other unmanned (Wheelon). They review the overall space program, and provide a setting for the various chapters that follow.

Part II consists of six essays that examine the policy context in which our space program is embedded; a context not often analyzed in discussions of space policy. Typically this context is considered a somewhat mysterious boundary condition. The essays address the overall budget situation (Macauley); the Congressional budget process (Telson); decision making in the Federal government and in industry (Fellows and Guastaferro, respectively); and how programs are affected by planning and management processes (Faget and Coleman). The last two essays – by Faget who looks at the lengthening development times of manned programs, and by Coleman who looks at why unmanned science missions have grown in size and complexity – provide a transition to the next part which discusses programs more specifically.

Part III consists of five essays that address science, applications, station, and shuttle programs. McCray and Stern examine the policies that drive the space science program and recommend several changes. Stern and Habegger examine a particular issue – why and with what result have we switched largely from multi-mission to single-mission space science programs. Webster looks at NASA's Earth Observing

System (a multi-mission program by the way) which will support both fundamental earth science research and also policy discussions related to global change. Finally two manned programs are studied: Brunner et al look at the future of the Station program while Pielke and Byerly consider the history of the Shuttle program as a caution for new launch system developments.

Part IV contains two concluding chapters. Limerick applies the lessons of the real American frontier to the real space program, in contrast to the usual comparison of a mythological frontier experience to a hoped-for space program. The final chapter attempts to articulate the overall message of the book. That chapter also discusses what more needs to be done, in hope of stimulating additional research and writing to fill gaps.

Each chapter is self-contained and the reader is welcome to read as desired: The arrangement is logical but the order is not critical.

Notes

1. R. Byerly, Ed., *Space Policy Reconsidered*, Westview, Boulder, 1989.
2. Ibid., p. 1ff.
3. A. Roland in ibid., Chapter 2, "Barnstorming in Space: The Rise and Fall of the Romantic Era of Spaceflight, 1957-1986," p. 33ff.
4. See for example, Task Group on Priorities in Space Research f Phase I, Space Studies Board, National Research Council, *Setting Priorities for Space Research*, National Academy Press, Washington, D.C., 1992. The Task Force was chaired by John Dutton.
5. On July 19, 1989, the twentieth anniversary of the first landing on the Moon, President Bush announced that the U.S. would return to the Moon and go on to Mars. He asked NASA to study how this could be done and report to him in 90 days. The resulting report was entitled *Report of the 90-Day Study on Human Exploration of the Moon and Mars*, NASA HQ, Washington, D.C., November, 1989.
6. Almost no funds were appropriated for the program in the subsequent budget cycle. Recently NASA has revised its approach and seems to be on a more reasonable track.
7. *Report of the Advisory Committee on the Future of the U.S. Space Program*, N. Augustine, Chair, USGPO Washington, D.C., 1990.
8. See note 4.
9. *Report of the Earth Observing System (EOS) Engineering Review Committee*, E. Frieman, Chair, September, 1991. This review was established by NASA in response to the National Space Council. It has not been published but will be reprinted in Subcommittee on Space, Committee on Science, Space, and Technology, U.S. House of Representatives, "Hearing; Earth Observing System Engineering Review," Sept. 19, 1991. Serial No. 72, USGPO, Washington, D.C.

10. B. Moore, III, Chair, Space Science and Applications Advisory Committee, letter of August 27, 1991 to L. Fisk, Associate Administrator for Space Science and Applications, NASA HQ, Washington, DC. This letter was the report of a five day retreat to plan space science missions.

11. Augustine report, (see note 7), p. 2

12. (a) NASA Advisory Council, *Report of the Task Force on Issues of a Mixed Fleet*, NASA, Washington, D.C., March 11, 1987. (b) NASA Advisory Council, *Report of the Task Force on Space Transportation*, NASA, Washington, D.C., Dec. 1989. Both task forces were chaired by J. Welch.

13. Resistance takes many forms. Consider the Augustine recommendation that science have "first priority." This has been cleverly, but not wisely, resisted by publicly agreeing with it and pretending that, indeed, science has always been a top priority. This tactic is clever because it disarms fainthearted critics but it is unwise because it is untrue, which further undermines NASA's credibility, and more importantly because it is very likely not what the Nation wants from its space program. That is, it seems clear (to this writer at least) that manned spaceflight is truly the top priority for our civil space program in NASA, in Congress, and among the public. Science is interesting (as in curious) but not a driver of programs or budgets. Had NASA constructively resisted that Augustine recommendation and stimulated an informed debate, it very likely would have prevailed, leading to a program aligned with its public goals. Instead there is another disconnect, another reason for citizens to distrust the Federal government in general and the management of the space agency in particular.

Evidence that Congress supports manned spaceflight is seen in the vote on June 6, 1991 to restore funding for Space Station. Some of the debate preceding that vote was framed in terms of manned programs versus science. This vote was on an amendment to the appropriations bill providing funding for NASA. The bill as reported by the Appropriations Committee did not contain any funding for Space Station, which the Administration had requested. Representative Chapman offered an amendment to restore funding and it passed by a recorded vote of 240 to 173. See *Congressional Record*, June 6, 1991, p. H4021.

14. For example, although NASA has been encouraged not to use the Shuttle to launch unmanned satellites that could be launched on an expendable launch vehicle (ELV), it still uses the Shuttle to launch many of its own satellites, because the Shuttle costs are built into the budget (i.e. Shuttle will fly in any case) but an ELV would have to be purchased on the market.

15. R. Brunner, *Performance as Promised: Restructuring the Civil Space Program*, Paper presented at the Thirteenth Annual Research Conference, Association for Public Policy Analysis and Management, October 25, 1991, Bethesda, MD.

16. That is, many supporters of manned spaceflight and human exploration of space would argue that the *particular* program, i.e. Space Station Freedom, is going to be a negative factor, in essence absorbing a great deal of attention, talent, funding, and political capital with minimal payoff for exploration.

Part One
OVERVIEW

The following two papers taken together provide an overview of the U. S. Space Program and thus a context for the papers that follow. Wheelon discusses the unmanned program and Roland the manned.

Wheelon looks ahead at how the changing world political situation will affect the unmanned efforts which have practical applications. He foresees continued practical use of space.

Roland sees two possible paths for the manned program: In the first option the manned program would be funded, perhaps to go to Mars, as a way to cushion the fall of Department of Defense spending in the aerospace industry. The second path would see a retrenchment of manned spaceflight for two reasons; first the absence of its principal rationale, i.e. the need for Cold War prestige, and, second, the pressure of constricted budgets.

Chapter 2
THE FUTURE OF THE UNMANNED SPACE PROGRAM

Albert D. Wheelon

SPACE IS A PLACE, NOT A MISSION.
—H. YORK

Introduction

The remarkable changes taking place in the world will inevitably affect the American space program. The purpose of this paper is to explore how these developments will influence the unmanned component of our space program. This portion is primarily the military space activity which will consume about eighteen billion (1990) dollars this year. It is also the unmanned scientific missions of NASA whose budget this year is approximately three to four billion dollars. Its third component is the commercial market for communication satellites and their launch vehicles, in which two billion dollars of private funds are invested each year. These quite different activities are influenced by different forces, not all of which are making headlines. Let us begin with two of the less obvious forces.

Fiber optic cables and communication satellites are competing vigorously for a growing market. This competition is affected as much by regulatory and trade policy as it is by technology. A wave of deregulation and competition is sweeping the world of telecommunications. The conflict between "cream skimmers" and rate averaging monopolies is a central issue. So too is the question of access to launch vehicles. Satellite operators try to reduce their capital investment by buying the cheapest launch vehicle so as to compete better with terrestrial facilities, while domestic launch vehicle providers try to discourage use of "subsidized" foreign rockets. How such issues play out will influence the survival of the first and only commercial success of space so far.

Aerospace companies will be strongly influenced by cuts in the defense budget. They are already in weakened financial condition. Changes in defense procurement policies require that they now provide considerable amounts of working capital, where previously they provided almost none. When these companies go to the financial markets to raise working capital in the present environment, they find it difficult to do so because they are not viewed as good investments. This progressive weakness may make them reluctant to compete for fixed priced programs on the international market. Instead they may increasingly seek shelter in the declining amount of cost-plus programs from DOD and NASA. This possibility is aggravated by the reluctance of the administration to adopt an industrial policy that stimulates or shelters key technical industries and by the embarrassment of NASA's lone attempt to support American comsat technology via the Advanced Communication Technology Satellite program.

The Cold War is receding and with it the need for some space systems. The Warsaw Pact has effectively ceased to exist and the requirement to monitor its military activities has diminished. The Soviet Union is becoming more accessible. The remarkable space systems that were once developed to expose their activities are being replaced by on-site inspection procedures, force reductions, and the opportunity to see on the ground that which we once could only see from space. Because space systems are large and expensive, reduction of the defense budget will inevitably squeeze these once sacred programs.

A favorable development has begun with the creation of a staff mechanism in the White House for making space policy: the National Space Council. It is clear that the initiative to send men back to the Moon and on Mars was the President's concept. The search for imaginative ways to do these missions has been led by the Vice President. The Space Council has raised the important issue of whether we should build a few giant spacecraft, or more numerous small vehicles. They have been involved in reexamination of the remote Land Sensing Commercialization Act of 1984. All of this is welcome. It is clear that the Space Council fills an urgent need. One can only hope that a White House function that gives policy guidance to the Space Program will survive.

Foreign space activities are gaining capability and purpose. If they choose to do so, a united Europe could provide substantial resources to their military, scientific and commercial space program. The Japanese are establishing independence in space and plotting a course of their own. Europe and Japan have begun to resist American space leadership for reasons that are not hard to find. Limited flight opportunities for the billion dollar European Spacelab manned scientific capsule and the long delay of the Ulysses Solar-Polar spacecraft are sensitive issues. Recent problems with Shuttle launchings and Hubble mirrors do not help. Frequent redefinition of the space station has undermined our leadership. As it withers from its original objectives, the SDI program will sow doubts abroad among those who joined that cooperative effort. All told, we may face a considerable lack of confidence and rising independence among our friends and allies.

Overwhelming all these issues is the Federal budget deficit. To meet the targets mandated by law, Federally supported space programs will be reduced along with other activities. It will be difficult to generate new programs in this environment. The Soviet Union finds itself in even more difficult circumstances and is trying to sell its space technology on the world market. The Europeans may elect to spend their resources in rebuilding Eastern Europe rather than in space. Japan is still not convinced that space is a good place to invest heavily. Between those countries that *cannot* and those that *will not* invest large new sums in space, the next decade may well be a lean season.

Military Space

The space program supported by the Defense Department is half again as large as the NASA program. If one recognizes that the Military Program is entirely unmanned[1] and that the NASA activity is primarily manned, one finds that the Defense Program is about five times larger than the unmanned NASA activity. This means that the DOD is the most important customer for unmanned spacecraft and their launch vehicles. Its programs and its direction establish the tide for the rest of the country.[2] It is thus important to understand the military program and the forces that will influence it.

The apparent end of the Cold War is having a profound influence on defense policy and the defense budget. These issues are being debated in Washington and throughout the country. They are far from settled. However, one can make reasonable guesses about how the budget will look in five years. My view is that the level then will be approximately two hundred billion (1990) dollars, down one-third from today's level. Only small forces will remain in Europe and none in Korea. Our military will be based almost entirely in the United States with very few overseas bases. We shall be concerned about Third World trouble-makers as much as the Soviet Union, and our military focus will cover the entire world.

If this vision is correct, we shall want to know a good deal about many countries. Our intelligence gathering activities will change from the present concentration on the Soviet Bloc to the world at large. American forces that may be suddenly deployed overseas will need to be supported with weather data, communications, and navigation services anywhere in the world. On the other hand, space programs represent a noticeable share of the defense budget and they will inevitably have to contract with the budget. To see how this squeeze may take place, let us examine the major areas of military space.

Communications

The military communication satellites that provide UHF service to mobile forces will become more important, because they can support operations anywhere without overseas bases. The present system (Fleetsatcom) will be succeeded by one with more capacity (UHF Follow-on) and these types of spacecraft will be operated into the indefinite future. This service will represent a steady demand for satellites and launch vehicles.

The trunking satellites (DSCS) that provide worldwide service to military bases and embassies will continue, if only because the Government frequencies represent a unique military asset that is separate from the crowded commercial band. The current trend to complicate these DSCS spacecraft by adding on board jamming protection and nuclear hardening will be gradually reversed. Satellites that optimize communication capability in contrast to sophistication will be selected in the same way that the military has turned to commercial computers and other products for many of its needs.

The Milstar system to provide a sophisticated new communication satellite service for the nuclear forces is faltering now in the face of escalating costs and the declining threat of global nuclear war. This program is being reoriented and may end. In any case, we will return to the traditional ways of controlling the much smaller nuclear forces that will result from START reductions.

Navigation

The global positioning system being built by the Air Force to serve national needs is a good system with a clear purpose. It provides high accuracy to the military and a lesser accuracy to civilian users. It is an important system to continue because it can support worldwide deployments. The USSR has fielded a similar system (Glonass) that is almost interchangeable with GPS. They have described their satellite and user terminal in detail to the International Civil Aviation Organization and urged its use. The FAA has held bilateral discussions with the Glonass program, but the DOD has opposed such cooperation. The military's reluctance to depend on other people's facilities is understandable, both here and in the USSR.

With continuing budget tightening, the logic for facility sharing will become too strong to ignore. Several solutions are possible. One is for a worldwide system based on the Intelsat model in which each country bears a share of the cost that is proportional to its use of the system. Another possibility is for the USA and USSR to maintain the worldwide system jointly, with each party providing some of the spacecraft.[3] A third possibility is for the USA to act as executive agent for the world, collecting contribution shares from using nations and building the system. This third possibility would require a good deal of trust in us by the other nations. The important long term question is whether the military will need the high accuracy capability. If it does not, a cooperative civil system based on the civilian accuracy should be established and its costs shared equitably among using nations.

Surveillance

This is an extremely difficult subject to discuss because of the special security in which it has always been wrapped. One can say a few important things. The program was mentioned by Presidents Johnson and Nixon, and formally announced by President Carter. It is specifically identified as "National Technical Means" (NTM) in a number of arms control agreements. This program is widely understood to have been a principal stabilizing influence in the Cold War, especially during the strategic nuclear buildup phase. But the world in which it operates is changing.

The Warsaw Pact has ceased to exist. Germany is reunited. Democratically elected governments have emerged in almost every Eastern European country, and one can travel freely throughout the region. In the Soviet Union, the situation is changing rapidly and will evolve further. The Soviet government now offers for sale technical data that was once sought by the most dangerous clandestine means. A large amount of defense information is regularly exchanged to establish data baselines for arms control negotiations. On-site inspection of weapons deployment and destruction is a feature of the INF, START and Nuclear Test Ban Treaties. We are rapidly substituting direct verification for indirect observation.

The Soviet space programs that monitor American military activities have been working against an open society. Because of this, they have not found it necessary to build NTM systems as capable or diverse as our own. If the Soviets evolve into an open society, *we* may not need systems that are as elaborate as those we now deploy. This judgment is contrary to current wisdom, which holds that we should improve our eyesight and hearing as we reduce our military forces. If we continue to face an intransigent, secretive and heavily armed adversary that is the correct strategy. If we evolve into a new situation however, it is important to devise a new strategy. In any case, budget pressures will force rethinking simply because NTM programs are so large and expensive.

Intelligence gathering will continue to be important, but the techniques will change as access is eased. The targets of collection will change too, going from a major focus on the Soviet Bloc to concern for many areas and many problems. Advanced weapons in the hands of Third World countries are a source of concern to both the US and USSR. Space systems will continue to play key roles in monitoring such situations, but they may be different systems from those we use today. Classical espionage will supply much of the needed information on such countries.[4] The long term prospect is for smaller NTM systems better adapted to a more open society and to Third World problems.

Missile Warning

Space systems that warn when ballistic missiles are launched have been a reality for twenty years. They have provided security against surprise attack for both the US and USSR. They have been the great stabilizers in the Cold War, giving confidence to both parties and preventing missile launch through ignorance, fear or confusion. And they are here to stay. This monitoring will be needed indefinitely since sizeable nuclear missile forces will remain on each side, even after two rounds of START reductions. It is unrealistic to expect that the US would rely on a Soviet system or they on ours, and two independent systems will continue.

A new threat is emerging. It is the seventeen smaller nations that now have ballistic missiles. The countries are mostly on the periphery of the Soviet Union and cause understandable anxiety there. An additional concern is that these nations probably do not have the discipline, the tradition, or the command, control and communication systems that have characterized US and USSR strategic forces. The Soviet missile warning systems are technically limited and cover primarily American missile silos. By contrast, the US systems provide global coverage and there is no clear need to upgrade our present system.

Space Defense

The Strategic Defense Initiative of President Reagan has largely run its course. It served a very good purpose if it brought the USSR back to the INF bargaining table and encouraged the negotiation of various nuclear force treaties. The technology developed by the SDI program has been quite remarkable and will find wide application. However, the original goal of providing a wide area defense now seems beyond our reach. To field such a system would involve bending or breaking the ABM Treaty. Moreover, Defense budget pressures will resist the deployment in space of an SDI system because of its cost: $50 to 100 billion. The SDI has only weak support among the military services who view it as a budget competitor to their own programs. There is little support in Congress for deployment. Unless a new threat emerges, the SDI program will remain a technology development program at the level of several billion dollars a year. As such, it will support space technology, but no large procurements of new satellites and launch vehicles.

Communication Satellites

Communication satellites are the first and most important commercial application of space technology. Certainly we are having trouble finding a second. Perhaps the first is enough. After all, the world has flown some 100 commercial satellites and is readying 60 more. They are all unmanned. In today's dollars, this represents an investment of private funds in spacecraft and launch vehicles of $13 billion over the last 20 years. These satellites generate some $3.5

billion in revenues each year. In 1985 alone, 18 satellites were flown for 12 different commercial customers who invested $1.35 billion in the spacecraft and launchers. It is respectable business.

International Service

Intelsat is an international telecommunications cooperative established by treaty among 119 nations. It was planned to be a monopoly that provides virtually all international voice, data and video traffic that goes by space. Intelsat is compelled to serve all members at the same rates. This rate averaging means that Intelsat must provide ten circuits to a developing country at the same unit charge as it provides 1,000 circuits to major users. The economic reality is that large users subsidize small users. This was understood and endorsed by the nations who joined Intelsat, but it represents a very substantial economic burden for Intelsat as it faces competing transmissions systems. Undersea fiber cables are Intelsat's most formidable competitors,[5] primarily because they only connect points with large traffic needs. Cables do not serve the widely dispersed, small users. By their very nature, cables are cream skimmers.

Most telephony and data will move to fiber cable on those routes where it is available. Major path satellite capacity may be retained as a result of policy decisions to subsidize small users or to provide emergency restoral when cables fail. Intelsat will continue to serve the small users who do not have cable service. If this were the only service Intelsat provides, it would be an increasingly expensive one.

However, satellites enjoy a unique advantage in transmitting television from one point to many points. In addition, time delay is irrelevant. Broadcast of major events is growing rapidly and transmission of television from one country to another for rebroadcast is growing even more rapidly. Video distribution will become the major revenue generator for Intelsat in the next decade.

To respond further to cable competition, Intelsat will be forced to specialize and particularize the spacecraft that serve different ocean basins. The traditional practice of buying all satellites to the highest expected demand[6] will be abandoned as more and more Intelsat members are deregulated. Access charges by member nations will be progressively reduced to meet price competition. Direct access to Intelsat spacecraft by user terminals will grow rapidly and gateway terminals will become obsolete.

In the Communications Satellite Act of 1962, the United States consciously decided to donate its comsat technology to help found

Intelsat. Only America had the technology to build communication satellites and the rockets to place them in stationary orbit. We abandoned that powerful monopoly for the larger objective of binding the world together with satellite communications. At once, it was a generous, successful act of technology sharing and a diplomatic coup.

NASA played a crucial early role in supporting communication satellite technology, but dropped out in 1973 when its budget was tightly constrained and it appeared that private industry could carry the development load. Intelsat has emerged as the principal supporter of communications satellite technology because its unique service demands have pushed it to develop new technology. For almost 15 years Intelsat has pioneered communication satellite technology. It was the first to develop an operational synchronous satellite (Intelsat I), the first to develop a channelized repeater (Intelsat IV), the first to exploit spatial reuse of frequencies (Intelsat IV-A), the first to combine spatial and polarization frequency reuse (Intelsat V), and the first to use satellite switched time division multiple access (Intelsat VI). Intelsat has spent three billion dollars in developing this technology, most of it with American firms.

Mobile Service

Radio communication has always provided a unique resource for mobile users. Satellite communications is only the most recent and powerful technology to provide mobile service. Maritime satellite service is now provided by the international cooperative Inmarsat which serves more than 6,000 ships. Its satellites and launch vehicles have been built both by Europe and America. It is a natural monopoly that responds to a demonstrated need and willingness to pay for service. Inmarsat growth will be limited primarily by the number of ships than can justify an onboard terminal. Future growth will depend on serving the much larger number of very small vessels that are not terminal-equipped today. Given the technical realities of limited bandwidth, raw satellite power and antenna directivity, I believe that such vessels will be served by data channels, and that Inmarsat will continue its steady growth.

Aeronautical satellite service shares a frequency assignment at L band with the maritime service. Inmarsat has started to provide a shared aeronautical and maritime service. This effort will succeed because the established maritime traffic will support the satellites in orbit and the aeronautical service can be offered on an incremental basis.

Land mobile service by satellite is not yet a reality. Early demonstrations by NASA showed what the military knows: UHF is the right frequency because the user terminals can be omnidirectional. The civilian UHF frequencies were assigned to fire and police users, however, and land mobile telephone service was pushed up to L band. Inmarsat has offered to provide this service with its existing spacecraft and is making some progress. Several national and cooperative programs are trying to do the same thing. A novel approach was recently proposed by Motorola, whose Iridium system would use seventy-seven small spacecraft moving in low earth orbit. Each would act like an upside-down cellular telephone base station (with constantly changing roles and intersatellite trunking) to provide worldwide service. If this concept is valid economically, it will have a profound effect on the launch vehicle and spacecraft markets for Lightsats.

Domestic Service

No provision was made in the 1962 Communications satellite Act for purely domestic systems. No one seriously considered the technical possibility that a satellite's beam could be focused narrowly enough to cover a single country. In the mid 1960s, engineers began to see how to stabilize large antennas and form sculptured beams that could compress all the satellite's power into a small cone covering the United States or Canada or Australia. Because the power density of a focused beam would be substantially higher than an earth coverage beam, the ground stations could be smaller and cheaper. This meant that satellites could compete with terrestrial networks on an economic basis. Moreover, satellites seemed to offer a unique advantage for television distribution with their natural one-point to many-point broadcast capability.

The Federal Communications Commission invited applications under the Open Skies Policy in 1970 and eight concerns promptly requested permission to establish domestic satellite systems. The first generation domestic satellites showed that a satellite could quickly establish new networks and new services. With that realization, the logical basis for a network monopoly disappeared. It was the beginning of the deregulation trend that is now sweeping our telecommunication industry – and the world. The Open Skies policy decision was to have a profound influence on the economics of communications. It would unleash a rush of new technology. It would create a marketplace characterized by multiple suppliers and multiple buyers.

What was not recognized was that those applicants were being given an opportunity to fail as well as to succeed. Open Skies meant the end of regulated monopolies in which earnings were assured as a percentage of the investment base by periodic tariff adjustment. Rates would now be set by competition. Three large firms that received FCC grants failed to proceed: Ford, Federal Express and Martin Marietta. Of the eight American firms that originally operated comsats over the US, four have gone out of business, been absorbed, or are withdrawing: Satellite Business Systems, Western Union, Southern Pacific Communication and American Satellite. Only four carriers seem committed to provide continuing satellite service for the United States: American Telephone & Telegraph, General Electric, General Telephone and Hughes Communications. The service they provide is primarily data and television distribution. Very little voice traffic now goes over satellite. ATT believes that better voice service is provided with terrestrial links because it has no time delay. It has shifted its large traffic base to microwave links and fiber optic cables. I believe that this is generally the way the domestic market will grow in the next decade: video on satellites and voice on terrestrial links.

A new phenomenon is emerging, largely unnoticed. Private networks using high frequency (i.e. K_u band) satellite capacity and very small terminals are being established. Major corporations are building and enlarging such systems. Their private networks bypass the regional Bell operating companies which are regulated monopolies. By avoiding the local operating companies they avoid the high charges imposed on incoming and outgoing traffic. In doing so, they undercut local rate averaging in the same way that satellites first challenged the long distance system.

National satellite communication systems have been established in nine countries besides the United States and the Soviet Union. Canada was the first to build a national system. Their Anik series has provided satellite service to the far north and to the entire country. A dozen spacecraft have been built in this program, primarily by American firms. Indonesia was another pioneer in satellite communications. With a weak terrestrial network, they sought to unify the country culturally and economically with comsats. A satellite system is a particularly powerful solution for a country divided into 3,000 inhabited islands. Using five American built satellites and rockets they established high capacity countrywide communications beginning in 1976.

Australia has also established a national communication satellite system. Their primary motive was to provide television and telephone service to the interior regions, since the terrestrial system runs primarily along the populated coastline. They chose American technology. Three satellites have been launched and two higher power versions are now being built. Mexico purchased four hybrid communication satellites from American firms and flew them on the shuttle. They provide wideband communications capability that supplements a weak terrestrial system. India has procured four American built satellites that provide a few direct broadcast channels, weather pictures and a number of telecommunication channels.

Brazil also has its population on the perimeter and its resources in the interior. After a decade of study, they purchased two satellites from Canada and established good communications in the interior. Theirs is the only satellite system in South America, but even so it is lightly loaded. Brazil has recently ordered two follow-on satellites from an American company. China has built and launched its own domestic communication satellites. France decided to establish a communication satellite system that would serve both military and civilian needs. Three Telecom spacecraft have been built by French firms and provide both metropolitan and some overseas service. Japan has built a succession of experimental communication satellites over the last twenty years. They have been small low capacity satellites because they were designed for launching on Japanese rockets. When the telephone system was deregulated in 1985, two domestic satellite service companies were authorized to establish high capacity satellite systems. Both turned to American firms for large spacecraft.

This accounting does not include three regional systems. Arabsat operates two satellites for the Arab world and Eutelsat five spacecraft that serve Europe. Both series were built in Europe. A new shared system in the Far East was established this year using an American spacecraft and a Chinese launch vehicle.

These overseas domestic and regional systems represent the commercial *free market* for which American, European, Chinese and Russian providers compete. It is usually characterized by international open tenders, with fixed price contracts for both launch vehicles and spacecraft. On the other hand, trade offsets, launch insurance and government-backed financing are increasingly important in these competitions.

Direct Broadcast

Broadcast of television directly to the home by satellite (DBS) was an early dream of space technologists. It was opposed by the television networks and their affiliated stations who feared competition for their lucrative franchises. NASA and the Canadians managed to fly an experimental DBS spacecraft in 1976 and it clarified the technical questions regarding attenuation and required power levels[7]. However, commercial DBS has taken almost twenty years to emerge. In 1982, eight American companies filed with the FCC to build such systems and were granted licenses. Only the Comsat Corporation actually proceeded to build satellites and buy launch vehicles, but it canceled the program shortly before the first planned launch and suffered a heavy financial loss.

One might ask: "What went wrong?" In framing the applicable standards, the responsible international body (the CCIR) ignored the engineering results of the NASA/Canadian experiment, setting both unrealistic and unnecessary technical objectives. In addition, the matter was pursued too long by engineers who did not appreciate the dominant influence of economics. Direct broadcast by satellite is primarily a business, not a technology. It is basically show business. The major networks spend a billion dollars annually for programming. To buy and install two million terminals for a minimum viewing audience requires an initial investment of about $1 billion dollars. Against those investments, the cost of satellite transmission service is inconsequential. It is only important that it work.

Commercial DBS systems are already providing service *abroad*. Satellites built for licensees in England and Luxembourg now provide television distribution to much of Europe. The Australian domestic satellite provides DBS service directly to rural areas. As other nations and *linguistic groupings* move to exploit this unique opportunity, it is likely that the market for DBS spacecraft and launch vehicles will become significant.

Earth Observation

The earth's surface and its atmosphere are routinely observed from space. Earth observation is characterized by multiple players and different approaches. Weather patterns in the atmosphere are observed by low altitude polar satellites and by synchronous spacecraft provided by six nations. The data gathered by these satellites is *shared without charge* as part of the international cooperative effort to predict the world's weather. The *surface* of the earth is observed by low altitude polar satellites operated by the USA, USSR and France. Each *sells* their data to countries, companies, and individuals for a wide variety of purposes. This is a good time to reconsider the activities from a global perspective because, quite literally, that is what the systems seek. In doing so, we can ask some hard questions about the American component programs and how they may evolve.

Landsat

A civilian program to observe the earth's surface was begun in 1970 by NASA and later operated by NOAA. Five spacecraft have been flown and a sixth is under construction. By observing the earth in different spectral bands and by measuring the absolute energy in each band very accurately, one can measure and sometimes predict many important things: future agricultural yields, the location of oil and mineral deposits, snow and water runoffs, pollution levels in rivers, lakes and the ocean, the location of underground water, the extent of forest blight and the destruction caused, etc. This data is sold to exploration companies and to scientists trying to understand the environment. It has been provided to foreign governments, fifteen of which have built their own Landsat ground stations (at $15 million each) so as to receive the data in real time directly from the satellites. The service was so successful that the French built a similar system (SPOT) with improved resolution. The Soviets have recently begun to sell photographs of the earth's surface that are probably derived from their surveillance satellites.

It is natural to provide this global data through an international cooperative like Intelsat. A single entity could develop the needed technology, procure spacecraft and launch vehicles, operate the space system and distribute the data. If this were done, no single country would have to bear the entire cost of developing the technology and maintaining the spacecraft in orbit. The cost of duplicative and competing systems would be avoided. The data format would be standardized and data access roles negotiated. Improved spacecraft sparing and higher confidence of data continuity would be provided. None of those virtues is present in the systems now flying. International discussions have begun to create such a cooperative arrangement and the United States should actively encourage the move.

The American service for observing the earth's surface (Landsat) has fallen into deep trouble. Under the Landsat Remote Sensing Commercialization Act of 1984, the Federal government *gave* the Landsat spacecraft and the accumulated data bank to a private company. Eosat is free to charge whatever it wishes for this data. It has the right to collect $600,000 each year from the fourteen countries that operate Landsat ground stations. Eosat has declined to invest its own funds in follow-on spacecraft, with the result that a service gap of two years or more will occur after Landsat 5 fails. The Administration has recently testified to Congress that the commercialization of Landsat is a failed policy and that a new solution is required.

The current national programs could merge naturally into an international consortium. When Inmarsat was created, the two competitive carriers (Comsat and the European Space Agency) then providing satellite maritime service leased their existing satellite to the new consortium while it prepared to build its own. In this way, Inmarsat got off to a fast start, while the existing providers could depreciate their investments. The analogy is quite clear: Landsat, SPOT and Soyuzkarta could provide data to a new consortium until it could build its own spacecraft, data bank and distribution policies.

Weather Service

If an international consortium makes sense for Landsat service, it makes even more sense for weather satellites. No nation now operates its weather service as a profit making business[8] and Congress has legislated that American weather satellites *shall not* be commercialized. A plan to deploy five synchronous weather satellites was devised by international agreement among national weather services. The plan has suffered in its implementation. The USSR has not yet flown its promised spacecraft over the Indian ocean. For long periods, the European spacecraft were unspared and sometimes not available. The two American satellites have suffered early failures: only one is now on station and it may fail soon, while the planned replacement spacecraft are long delayed.

These problems could be corrected with an international cooperative. A consortium operating five identical satellites can afford in-orbit sparing, whereas single operators cannot. A single buyer can drive better bargains for spacecraft and launch vehicle procurements on larger lots. It can support the advanced technology that is needed. Buying the spacecraft to a single standard will reduce units costs and improve reliability. Such an arrangement would provide better coverage at lower cost to each participant. If the costs of the system were spread over all users, the cost of the service to the present providers would be *greatly* reduced.

The situation for low altitude weather satellites is a little different. NOAA and the Air Force both operate low altitude spacecraft that are very similar. Recurring efforts to merge the two systems have failed, and it is less likely that the military would be willing to defer to an international system. On the other hand, if an international consortium were created to provide Landsat coverage, it would make scientific sense to merge the low altitude civilian weather service with it. The two types of spacecraft fly in similar orbits and make complementary measurements.

Earth Observation System (EOS)

NASA plans to develop and operate a large satellite system that will observe environment status and changes on the earth. It is an expensive program[9] and will consume almost all the funds the Unites States has committed to environmental preservation. The program is fundamentally flawed in two ways. It puts too many environmental instruments on a single polar spacecraft. This strategy is a legacy of the manned polar space station dream that is now dead. The plan to put all our eggs in one basket is now justified on the basis that simultaneous measurements are required and so all the instruments must ride on the same carriage. Close scrutiny shows that this is not so.[10] It is apparent that the necessary measurements can be performed more cheaply and reliably if they are carried on smaller, separate spacecraft.

The second problem with the EOS program is that it comes too late. If we follow its plan, we will not get our first environmental data from space until the end of the century. The problem is urgent today. The more sensible course is to fly some instruments as soon as possible, to learn from then and to design later experiments on the basis of that experience. This adaptive, evolutionary approach is normally followed in science. By waiting until the last instrument is available for a large complicated spacecraft, we delay and compound the effort. The present program is too expensive, too vulnerable to failures, and too late to respond to the urgent need for environmental understanding and action.

Scientific Missions

The manned space flight community has repeatedly urged scientists to fly their instruments on shuttle and the space station. Most of the scientific community has been reluctant to go along. Manned space flight launch dates are often very unpredictable: witness Challenger-induced delays. It costs far more to qualify experimental apparatus for manned space vehicles than unmanned. The need for very precise pointing of many scientific experiments means that the presence of moving astronauts and related life support can disturb the scientific measurements in significant ways. Inevitably, scientific objectives are subordinated on highly publicized manned missions. These are some of the reasons that most of the scientific community has a clear preference for doing their experiments with unmanned spacecraft.

Planetary Exploration

The last few years have been a remarkable period for planetary exploration. Voyager visited Neptune. Galileo was launched to Jupiter and Magellan to Venus. The European Ulysses spacecraft has been launched and will fly high over the sun. This clustering of events has stirred public awareness and excitement about the program to explore the solar system. However, the current rate of activity is misleading. Voyager was launched in 1977 and had a fixed date with Neptune. The launching of scientific missions was long delayed by the Shuttle accident and their flights this year were artificially bunched. It is clear that we will not maintain the current level of activity in the future.

The Mars Observer will be launched in 1992. The CRAF/Cassini program has now started to build two spacecraft: one to explore Saturn and the other to rendezvous with comets. They will be launched in 1995 and 1996, and will complete their missions as the century ends. The prospect is for a continuation of the modest pace of planetary exploration that we have seen for the past decade.

Two things could change this. If the space station or Shuttle runs into further trouble, funds could be taken from the planetary program as they were before. Another possibility is that the President's initiative to send people back to the Moon and on to Mars will go forward and stimulate an expanded unmanned reconnaissance effort as a precursor. Such a push would focus its effort almost exclusively on the Moon and Mars. While it would develop significant robotic technology in the process, it would postpone further the return missions to the inner and outer planets that are required.

Great Observatories

The first of the four great observatories has just flown, carrying with it great expectations. Frustration with the Hubble Space telescope's faulty mirror is now shared by NASA, the scientific community and the people who paid the bill. It is too early to estimate the impact this mistake will have on public and congressional confidence in NASA's ability to mount the other large astronomical missions. Yet to fly are the Gamma Ray Observatory in November, 1990, the Advanced X-ray Astronomical Facility in 1997 and the Space Infrared Telescope Facility in 1999. All will measure radiation that is normally screened out by the Earth's atmosphere.

The situation is different for optical astronomy. Optical scintillation in the atmosphere smears the image of a star by a tiny amount. In addition, backscattering of light from the Earth's surface tends to obscure light from the faintest stars. It was to overcome these two limitations that Hubble was built at a cost of $1.5 billion. In the meantime, the Keck telescope has been constructed with $100 million of foundation money on the Island of Hawaii and achieves almost the same performance level planned for Hubble. An active array telescope is being built by a European consortium in Chile that will exceed the performance of Hubble. The great cost difference plus the painful error in Hubble will raise searching questions about the wisdom of doing astronomy in space. Fortunately, there are good reasons for doing so. The Earth's atmosphere screens out many of the interesting electromagnetic waves that teach us the history of the universe. Infrared, ultraviolet, X-ray, millimeter waves are all absorbed by the atmosphere, and it is necessary to go into space to receive these signals. For this reason alone, we must continue the great observatory programs.

Launch Vehicles

The United States has reestablished a diversified launch vehicle capability, having reversed its previous policy of exclusive reliance on shuttle. We now have seven launch vehicles in service: Scout, Titan 2, Delta 2, Centaur 2, Titan 3 transitioning to Titan 4, and the Space Shuttle. Given the rate at which they can be built and launched, we can place 860,000 pounds in orbit each year. This number can be doubled if the launch rate is quickened and their performance enhanced by marginal improvements. We should ask how the present lifting capability compares with the demand. We launched 600,000 pounds in 1985 and 1986. We launched less in 1990. There is thus, a considerable margin between our need and our capability. If any of the rockets has a problem, satellites planned for it can be shifted to comparable launchers with some accommodations.

In view of this robust capability, one must ask why we should develop a new, large launch vehicle for space. There are three motivations. (1) The Strategic Defense Initiative program planned to launch large amounts of materiel into space to provide a shield against potential missile attacks. The material required for a deployed system was so large that it would have outgrown the lifting capability of the present fleet. For this reason, SDI became the primary sponsor of the Advanced Launch System (ALS). (2) As NASA has evolved its plan for building and resupplying the Space Station, it has become concerned that the Shuttle could not transport material fast enough to support the station. It has proposed that Shuttle technology be used to build an unmanned resupply vehicle called Shuttle C. (3) A third push comes from those who believe in creating new capabilities, independent of specific needs. This group believes in establishing space infrastructure in much the same way that we built railroads and highways a century ago.

Let us take these arguments in reverse order. The Russians have recently built the large Energia launch vehicle that can place 220,000 pounds or more into earth orbit. It flew successfully in 1987, but is now grounded for two years because there are no large payloads for it to lift and because it is so expensive. Soviet scientists have criticized Energia because they have little to fly on it since its development has drained funds from the satellites they might have built. In a zero sum budget game, new launch vehicles are usually supported at the expense of satellites they are supposed to lift. The creation of infrastructure launch vehicle capabilities is thus a decision to be made very carefully.

The Space Station may need additional lift capability. If so, that does not lead one automatically to Shuttle C or ALS. Additional cargo launches with existing American rockets could fill this need. So too could foreign launch vehicles. In an increasingly cooperative world, the Russians have offered the Proton and Energia launch vehicles on a commercial basis. The European Ariane rocket is available to the world and is regularly used by American satellite communications companies.

This brings one to SDI as a unique driver for a new large launch system. Whatever one's position on the technical and strategic merits of a missile shield, it is clear that SDI deployment faces high hurdles. In a rapidly tightening Defense budget, the SDI program is meeting growing resistance – both from Congress and within the Pentagon itself. SDI can only be deployed with a substantial modification of the ABM treaty and such a move would run counter to the flow of world events. This is not to say that SDI should not be pursued; that is a larger question. What it does say is that we ought to proceed very cautiously on a *major new launch vehicle* program that depends for its justification on SDI *deployment*. I am forced to the conclusion that valid requirements do not now exist that would support the development of a new large launch system. The existing fleet will meet the needs of the space program America can afford today. In response to this logic and to budget pressures, the Air Force withdrew from its joint ALS program with NASA, and the project continues within NASA in name only.

If we move forward on President Bush's proposal to send men back to the Moon and on to Mars, we will want to develop a large rocket specifically sized to do that job, just as we built the Saturn Five rocket to send Apollo astronauts to the Moon twenty years ago. In the meantime, we should initiate a vigorous rocket technology program for future launch systems. We have neglected this area in recent years. If we develop competitive technologies, we will be in a position to move confidently into a new launch vehicle when a compelling need arises. Both Europe and Japan are investing heavily ($5 billion) in such technology and aim to decrease launch costs by as much as 40%. We must do likewise or fall behind by 1995.

Commercial space projects have a wider choice of launch. Some are already using the Ariane from Europe. Many would like to use the Russian Proton and the Chinese Long March because their prices are significantly lower. American suppliers of rockets oppose this and have received some government support to erect protective trade barriers. This debate about foreign quotas and subsidies will be resolved in favor of a quasi-free market for launch vehicles. Each country will subsidize its rocket production with government orders, thereby allowing the companies to sell to others on an incremental basis. In this sense, each country will subsidize its industry in the only way that counts: with a continuing, predictable series of orders.

Conclusions

The future of unmanned space programs is characterized by considerable uncertainty and risk. At NASA, the combination of budget reductions and swelling costs for manned programs is likely to cause unfortunate reductions in unmanned scientific missions – as it has in the past. Europe and Japan will become progressively more independent and less willing to join collaborative efforts with an unpredictable NASA.

Space systems for observing the earth's surface and atmosphere will move from isolated national activities to international cooperatives patterned on Intelsat and Inmarsat. Commercial communication satellite systems will meet increasingly effective competition from terrestrial systems. Their future will be one of gradual growth and periodic system replacement.

Military space programs represent the largest part of our national effort. It is entirely unmanned. The objectives of those military programs will change in response to the remarkable events taking place in the world and as our national security policy evolves away from an exclusive concentration on the Soviet threat. The present mix of systems will need to change to support the new policy that emerges. Some systems will be dropped, some modified and a few added. In particular, the SDI program will move from an operational defense plan to a technology development activity.

Notes

1. The last military interest in manned space flight was the Manned Orbiting Laboratory that was canceled in 1969.
2. To verify this assertion, note that expendable launch vehicles (Titan, Delta, Centaur) exist only because of large DOD orders; NASA and commercial users buy such rockets on an incremental basis from the DOD production lines. Sustained development of important components like traveling wave tubes, thrusters, and solar cells have been funded by DOD, and to a lesser extent by commercial customers.
3. This number might be greater than half of the total number of satellites, so that each would be partially protected against a withdrawal of the other.
4. Note that the Israeli Mossad keeps good track of the world without a National Technical Means.

5. Some maintain that fiber cables are cheaper than comparable satellite capacity on a per annum basis. Comparisons of installation costs for TAT 8 and Intelsat 6 show that cables actually cost about 50% more per year of service. This raw transmission investment is rapidly overwhelmed by access charges and preferential loading effects. As a result, heavily loaded cables are economically preferable to satellites that must serve all nations at common rates.

6. A practice encouraged by rate-based economics.

7. The Japanese built a series of experimental DBS spacecraft without much success. They were joined in this frustration by a German/French consortium.

8. A few small value-added firms provide specialized forecasting using the US Weather Bureau's basic forecasts as their foundation.

9. Costing seventeen billion (1990) dollars before first launch in 1998 and thirty billion before the end of its data-taking lifetime.

10. C.D. Graves, P. Gottlieb & A. Rosen, *Coordinated Measurements for Earth Observation*, TRW Report, June '89.

Chapter 3
THE LONELY RACE TO MARS:
THE FUTURE OF MANNED SPACEFLIGHT

Alex Roland

WE DON'T NEED TO PROVE OURSELVES.
—C. JOHNSON[1]

Manned spaceflight is an artifact of the Cold War. So too, for that matter, is NASA. Now that the Cold War is over, the future of both is cloudy.

The Race Begins

The primary argument for manned spaceflight has always been prestige.[2] The enthusiasm to put men in space began after Sputnik, when the Soviet Union seemed to have stolen a march on the United States. There was strong sentiment then for a race, if not in the White House then at least in Congress and seemingly the public at large. Senate Majority Leader Lyndon Johnson believed that "a new era of history [had] dawned over the world." Soon he and others realized that the new era posed an unprecedented military threat to the United States, for the same rocket that could put a satellite in orbit could put a bomb on New York. The press swarmed into print and into the airwaves in what historian Walter A. McDougall has called "an aimless, agitated 'media riot.'" A space race was on, and the first man in orbit would win the next heat.[3]

President Eisenhower tried in vain to quell the mania. He brushed aside questions about a space race in his news conference of 9 October 1957, five days after the Sputnik surprise.[4] He felt that missile programs he had approved in 1955 answered the Soviet military challenge; there was no need to engage the Russians in a basketball game in space, as one of his aides characterized it. His newly appointed science adviser, James R. Killian, President of MIT, later explained why Eisenhower was advised not to get into a manned "space race" with the Soviets. "The really exciting discoveries in space can be realized better by instruments than by man," he said. "The Soviets . . . have used technology as an instrument of propaganda and power politics," but that was no reason for the United States to follow suit.[5] Surely Eisenhower was "not about to hock his jewels," as he told the National Security Council in 1960, to race the Russians to the moon.[6] His successor, John F. Kennedy, took a different view. He campaigned on the theme of getting America going again, as in a race, and he announced in his inaugural address that Americans would "pay any price, bear any burden, meet any hardship . . . to assure the survival and the success of liberty."[7] His rhetoric, however, was not sufficient to reverse the course of events. Cosmonaut Yuri Gagarin became the first man in space, completing almost a full orbit of the earth on 12 April 1961, more than three weeks before American Alan Shepard rode his Atlas rocket to a 15-minute suborbital flight down the Atlantic test range. The United States did not match the Gagarin achievement until the Mercury-Friendship 7 flight of John Glenn in February 1962. By then the Russians had sent cosmonaut Gherman Titov on a multi-orbit flight of 25 hours duration. All the first rounds of the space race went to the Russians.

Gagarin's triumph, the most important achievement, was compounded within a week by the disastrous failure of the Bay of Pigs invasion of Cuba. Kennedy sought desperately for some public gesture to restore American credibility and prestige. His science adviser, MIT President Jerome Wiesner, suggested a national commitment to desalinization of seawater.[8] His vice President, Lyndon Johnson, had a better sense of what was needed. Relying on advice from NASA, Johnson told Kennedy that the U.S. had a fair chance of beating the Soviets in a race to the moon. Within days Kennedy was standing before a joint session of Congress delivering a speech on "urgent national needs." Space was "not merely a race," he said, but he nonetheless challenged the country to "take a clearly leading role in space achievement. . . . For while we cannot guarantee that we shall one day be first, we can guarantee that any failure to make this effort will make us last."[9] He asked for a commitment to send an American to the moon and return him by the end of the decade.

To the surprise of Kennedy and most of his administration, congress bought the moon initiative. Even after the price tag was revealed, $20 - $25 billion (that is $91 - $114 billion in 1989 dollars), Congress didn't blink. Perhaps, as Kennedy had asserted in his inaugural, Americans really would "pay any price." They surely seemed to like the idea of a race to the moon.

The race did not last long. At first the Russians tried to keep up, riding the propaganda tiger unleashed by their early successes with Sputnik and Gagarin. But as American analysts had guessed, to go to the moon both countries would have to develop a new rocket of unprecedented power. This robbed the Soviets of their previous advantage — the huge ICBMs developed in the decade before Sputnik to send nuclear weapons to the U.S. These rockets far exceeded the power of the American missiles developed later and consequently with less lift weight, for they were designed to launch the early nuclear warheads, which were far heavier than the later devices based on lithium deuteride. Because an even more powerful rocket was needed to go to the moon, the race started even and the U.S. quickly took the lead. The Soviets released conflicting messages about whether they were racing at all. Throughout the 1960s they worked on a Saturn-class rocket, the so-called G booster, but it failed in three launch attempts and was abandoned. Other schemes for a manned moon mission proved equally futile.[10]

Technically the U.S. program was far more successful; but politically it soon lost steam. The initial U.S. target date for the moon landing had been 1967, for fear that the Soviets planned to stage their landing on the fiftieth anniversary of the Russian revolution. House Speaker John McCormack had convinced President Kennedy to leave himself some maneuvering room by promising a moon landing by the end of the decade, but insiders still believed that 1967 was the date to beat. As intelligence leaked out of the Soviet Union, it soon became apparent that the Russians were not going to make that date. Not long afterward it was obvious that the Russians weren't going to the moon at all, at least not anytime soon. How could the U.S. have a race if the Russians weren't competing? Even President Kennedy shocked the U.S. space community by suggesting in a U.N. speech in 1963 that we should go to the moon together with the Russians.[11] Critics who viewed the Apollo program as a "moondoggle" and worse were delighted, feeling that the funds for the crash program could be better spent on earth; space enthusiasts were appalled that their cherished goal could be abandoned so casually.[12]

The race proved to have a momentum of its own. The possibility that the moon mission would be abandoned or redirected evaporated on 22 November 1963, when President Kennedy was assassinated in Dallas, Texas. His successor, Lyndon Johnson, was too astute a politician to cancel the most visible and popular symbol of the New Frontier. What's more, Johnson was himself a space cadet of the first water, and NASA's Manned Spacecraft Center was in Houston because of the power and support of the Texas congressional delegation, of which Johnson had been dean before his election to the vice presidency.[13]

The die was cast. The United States completed its dash to the moon. By 1965 most of the money for the program was spent in any case, invested in research and development on the huge Saturn launch vehicle and the Apollo spacecraft and in the facilities at Houston and Cape Canaveral that would launch and control the manned missions. A consensus settled on Washington that the U.S. should play out this round of Cold War competition, but think long and hard about announcing another race with the Soviets. The consensus was strong enough to sustain the Apollo program through the dark days after the Apollo fire of 1967 and through seven of the projected nine trips to the moon.

The Great Middle Distance

Thereafter the two sides in the Cold War settled into a space race of a different sort. They played tortoise and hare. Perhaps the Russians could not beat the Americans to the moon, but they could keep flying in earth orbit indefinitely. They could repeat the Gagarin achievement over and over again. They soon learned that just being in the race was almost as good as winning. Manned spaceflight gave them the same advantage their huge arsenal of nuclear weapons did. It was a game only they and the U.S. could play. It was a superpower race, and they profited just by running.

The Soviets also discovered that even the tortoise won some prizes. They delighted in the chance to join the U.S. in the 1975 Apollo-Soyuz Test Project, for their cosmonauts met the veterans of the moon in earth orbit as equals.[14] They set out after records, not of speed but of endurance. Having sent the first satellite, the first dog, and the first man into space, the Soviets sent the first woman in 1963 and the first team in 1964. Through the first half of the 1960s they had the record for the most people and the most total hours in space. In 1978 they began carrying guest cosmonauts into orbit, enhancing their prestige in the communist world and rewarding those countries that toed the party line — successively Czechoslovakia, Poland, East Germany, Hungary, North Vietnam, Cuba, Mongolia, and Romania. Later they branched out to France, India, and Syria. These cosmonauts in orbit never won a battle in the Cold War, but they were a fleet in being and a force in the war of words.

The United States proceeded differently. The cost of completing its solo race to the moon came to be measured against other demands on the national treasury and psyche. In the late 1960s, protest over U.S. involvement in Vietnam combined with urban unrest, environmental concerns, and economic crisis. The result was a period of intense domestic turmoil. People took to the streets in protest. Urban violence exploded in Detroit, Washington, and other cities. Students seized universities and held them hostage to an inchoate but growing agenda of reform. Taxpayers and voters turned on their government, forcing Lyndon Johnson out of running for another term and threatening other incumbents as well. The country itself seemed to be coming apart at the seams.[15]

The Apollo landings distracted the nation for a while from its time of troubles. They were not enough, however, to reverse the slide in NASA's budget. Nor were they enough to stimulate any national consensus on what to do next in space. If the Russians weren't racing and we had achieved Kennedy's goal, then shouldn't we turn our attention and our resources to pressing ahead on earth?

Within NASA, however, the race and the future had become one. The agency had embraced the Kennedy rhetoric and designed a future around it. The Kennedy Space Center was originally planned to handle 50 Apollo launches a year.[16] Peak Apollo spending in 1964 at about $5.2 billion ($22.3 billion in 1989 dollars) seemed the appropriate steady state for the space agency. What was a one-time crash program for Congress and even for the president became in the eyes of NASA a level-of-effort benchmark. As the NASA budget declined in the late 1960s, even though the first moon landing was not yet, space enthusiasts sought formulae to reverse their waning fortunes.

In 1969, NASA embraced "the next logical step." Like the Russian space program, it was a doctrine designed at once to service and exploit the Cold War. If the Russian long suit was tortoise, then the U.S. long suit was hare. Instead of proceeding incrementally toward a range of objectives in space, NASA proposed to jump rabbit-like from one great goal to another. Now shuttle. Then space station. Then Mars. It was giant steps that had allowed us to overtake the

Russian lead in space. Giant steps would keep us ahead. What's more they would bring in the avalanche of funding that had smothered the agency in the early 1960s. If you already had the world's most powerful launch vehicle, the Saturn, then scrap it and start over.[17] This was how we had caught up and won the space race. The key to the future was running it over and over again.

The battle of the NASA budget has been fought in these terms ever since. Advocates of a large space program, especially a large manned program, see the early Apollo years as the model. They want the U.S. to once again spend .78% of GNP or 4.44% of the federal budget on space as it did in 1965. This compares with the .2% of GNP or .9% of the national budget the government now spends on NASA. Instead of a $12.3 billion budget in fiscal year 1990, NASA would have a budget of about $60 billion.

What happened instead is that NASA's budget declined steadily after 1965, bottoming out in fiscal year 1973, the first year of the second Nixon administration and the last year to witness men on the moon. When Nixon first came into office, he rejected the most ambitious proposals of NASA and told the agency that it could expect a steady-state budget of about $3.5 billion 1969 dollars through his tenure. If the agency wanted to pursue the shuttle under that ceiling, Nixon would allow it. But there would be no Apollo-like crash programs, no huge infusions of new funding. As it turned out NASA's budget remained remarkably stable through the period of shuttle development, not varying in any year by more than 10% above or below the 1973 figure in real dollars. It was not until the infusion of funding after the Challenger accident that NASA's budget began rising significantly in real dollars.

The Cold War rhetoric in these years has been conducted over the indefinite article. Does the U.S. have to be *the* leader in space or is it enough to be a leader in space? To be the leader in space requires America to be first in everything, from communications satellites to space stations to men on Mars. To be a leader is far less demanding. In fact, since the space race is such an exclusive enterprise to begin with, virtually everyone who runs is a leader of some sort. Surely the Russians have predicated their continuing support of manned spaceflight on the belief that running is almost as good as winning. Some even came to argue after the Challenger accident that the Soviet Union had overtaken the U.S. in space, a victory of quantity over quality.[18] Of course a good bit of that was Cold War rhetoric and posturing for more funds, but it must have sounded just as sweet in the Kremlin as if it had been true.

Those who would revive or restart or sustain the space race reject the argument that we have won. Some space activities like reconnaissance satellites and communications satellites and scientific probes are their own reward; the payoff is in the function of the spacecraft. The space race, however, has no payoff beyond prestige. A victory in one heat achieves nothing unless you also win the next one. Going to the moon was fine, but that proves only that we used to be better than the Russians. Now we must prove it again by going to Mars. There was nothing on the moon that was worth returning for. Nor is there likely to be anything on Mars worth returning for. The rocket we built to go to the moon did not open up space to us; we scrapped it. The technology we build to go to Mars is not likely to be any more useful. The prize is in the prestige; the purse is filled with Tang.

Historical Analogs

But what is prestige worth now that the Cold War is over? The answer seems to be "less." It is surely not trivial in a world driven by economic competition more than by military competition. Being number one, establishing or retaining a reputation for cutting-edge technology and science, conducting high-visibility research and development projects – all these remain important in the post-Cold War world. And because manned spaceflight has traditionally been seen as a measure of all these factors, its significance will no doubt continue. The very fact that the European space agency and the Japanese are developing manned spacecraft suggest that the field of activity will be important in the future as it has been in the past.

But not *as* important. The countries pursuing manned spaceflight now are the very ones at the leading edge of the struggle for economic mastery. Practical return on investment seems to be more important now than at the height of the Cold War. In fact, one explanation for the space race of the Cold War was that its wastefulness and pointlessness were a measure of its political utility. Historical examples of this kind of phenomenon are not hard to find.

Pyramids in the ancient world had at least three functions. First, they were tombs for the pharaohs. They served this purpose admirably, though we might say in the modern world they were over-engineered. But according to some scholars, that was just the point: they were over-built on purpose. They rose in the desert as awe-inspiring monuments to the power of the state and its ability to waste incredible resources on otherwise pointless enterprises. Any state that had the material and manpower to build such useless symbols, surely had the wherewithal to crush its enemies, both within and without. Those who considered challenging the Egyptian state were enjoined by the pyramids to consider what they were getting themselves into.

Furthermore, the laborers who worked on these monuments were building more than tombs and symbols. They were building the Egyptian state and the sense of social and political cohesion essential to the survival of the state. The pyramids were built not sequentially but continuously. For generations in early Egyptian history, workers liberated by the completion of one pyramid were put to work on the next, moving from the small-scale peak on one to the labor-intensive base of another. For over a century, the common people of Egypt bore the financial and physical burden of erecting a great, empty monument to the state. When they were done, they were proud of their work, and they were bonded with their countrymen by the shared experience of an onerous and imposing job well done. The more they suffered, the greater their satisfaction.[19]

The same kind of community bonding no doubt infected all peoples on whose backs great civic monuments were built. We know that the medieval cathedrals had this effect on their benefactors, in spite of the fact that the poor were burdened not only with the physical labor but also with the taxes that supported the construction.[20] The space program, and manned spaceflight in particular, surely fit this mold: an enormous, expensive, inspiring technological artifact whose cost in labor, lives, and treasure exceeds its practical utility.

Thorstein Veblen, the iconoclastic economist of the early twentieth century, described this kind of behavior as "conspicuous consumption."[21] Individuals, seeking to establish their reputations within the community, will consume in such a style as to convince others that they are rich. Wearing tennis clothes, for example, the 1930s version of modern leisure suits, suggested to observers that (1) you belonged to a club, (2) you had time to practice a sport that required extensive training (perhaps even had tennis lessons), and (3) that you had plenty of free time, i.e. that you were independently wealthy, or at least did not have to work regular hours. The problem with this kind of public posturing, of course, was that impostors were forever aping the behavior of the truly wealthy, driving them to new forms of ostentation. At one time, you mimicked the wealthy by wearing expensive clothes. When others appeared in expensive clothes the wealthy began wearing shabby and informal clothes, because only the self-assured would dare appear in public poorly dressed. Now pre-worn and pre-washed designer clothes are manufactured and sold at great cost so that the wealthy can still spend lots of money trying to stay ahead of the interlopers.

No national consumption in the last thirty years has been more conspicuous than manned spaceflight. More and more astronauts and cosmonauts were sent into orbit with no apparent function beyond demonstrating that our nation or theirs had more resources to squander on this activity — what Veblen called "specialized consumption of goods as an evidence of pecuniary strength."[22] The symbolic peak was reached when Alan Shepard went golfing on the surface of the moon; all he needed was a crocodile on his space suit.

Another analog of this kind of undertaking is potlatch, the custom of some American Indians of the Pacific Northwest. When two chiefs compete for position or advantage, they resolve the issue by means short of combat. They give away or destroy personal possessions — not their antagonist's, but their own. The contest continues until one chief runs out of material or cries uncle. The winner is the one who loses the most.[23]

By that measure, the space race is a tie. The American and Soviet economies in 1990 are both in disarray, in part because of the $300 billion apiece the two countries have spent on the public space programs. Surely the space programs of the two countries are not responsible for the economic mess in which they find themselves, but they are manifestations nonetheless of the larger policies that brought them to their present pass. The huge nuclear arsenals, like pyramids in the desert, were symbols of the power of the two states to bend their resources to purposes of war, cold or otherwise. Wrecking the economies of the two countries required other policy commitments, such as rampant consumerism in the United States and hopelessly inefficient state planning in Russia, but the potlach of the Cold War helped.[24] Any examination of policies in the post-Cold War world will have to include the manned space programs of the two countries.

Nyet and Not Yet

The Russians have already begun their re-examination. Even before the Cold War ended in 1989, there seems to have been some discontent in the Russian manned space program. The space station MIR, which was to have been permanently inhabited, was abandoned for some months in 1989, and is to be inhabited only intermittently in the future. One reason for this appears to be the discontent of the cosmonauts themselves, who are no doubt finding the boredom of space oppressive. Another reason is obviously the cost coupled with the lack of payoff. Though it is often argued that the Russians are making great strides in space manufacturing, they as yet have very little to show for their years of investment. Compounding these problems that were already visible before 1989, perestroika seems to have targeted manned spaceflight as well, seeing it both as a Cold War artifact and as an inefficient drain on the economy.[25]

Though the U.S. economy has problems of its own, President Bush seems to be more reluctant to call off the space race. His infamous campaign pledge to allow "no new taxes" is now a sad joke, but he pushes on nonetheless with an ambitious manned space program. He has continued the policy of Ronald Reagan of increasing NASA's budget significantly since the Challenger accident, asking for a whopping 24% increase in the 1991 budget. Most of this is to go to manned spaceflight, to support the space station, a return trip to the moon, and manned expedition to Mars. The last project, which has been the long-range goal of NASA planning since Apollo, Bush proposes to achieve by 2019, half a century after the first moon landing. In other words, he is still racing. The Russians don't want to race, in fact they are even cool to the idea of cooperating with the U.S. in a Mars mission – no doubt because they see little reason to go if it is not a race. But George Bush and the other cold warriors enamored of this project have been racing for so long they seem not to be able to envision a world without it.

Whether Congress will any longer fund such an adventure remains to be seen. Congressional support for NASA in general and the manned space program in particular has remained strong through good times and bad. But time may show that the Challenger disaster marked a turning point. That accident and the institutional flaws revealed in the subsequent investigations undermined NASA's credibility as did no other event since the Apollo 204 fire of 1967. Congress continued to support NASA after Challenger, even to provide the substantial funding increases the agency said it needed to fix its problems. But Congress has nonetheless been skeptical, and it casts an increasingly jaundiced eye on the space station and the proposed Mars mission – in part because of their costs and in part because of a diminished confidence in NASA. The failure of the Hubble telescope and the grounding of the space shuttle fleet in the summer of 1990 compounded a trend that was already working against NASA. No matter what President Bush may choose to propose for manned spaceflight in the future, Congress is probably going to play a stronger role in shaping the program than it has in the past. Though the authorizing committes in both houses show no signs of taking any initiative, the approriations committees may well seek overall funding cutbacks that will preclude any race to Mars or even to a space station.

The Prospects

Two paths seem open to manned spaceflight in the United States. One is history redux. The defense budget is now falling, as it was in the early 1970s at the end of the Vietnam war. The aerospace industry is particularly hard hit, as it was then. President Bush may well conclude, as President Nixon did then, that the aerospace industry needs a make-work project, if not to sustain the industry at Cold War levels then at least to cushion its fall to an appropriate post-cold-war level. Since the aerospace industry has been the mainstay of our declining balance of trade, there are good reasons to believe that it deserves government support. Especially is this true in light of the increased competition that can be expected from a liberated and economically united Europe, already a powerful force in the international aerospace market. Thus President Bush might propose major increases in the NASA budget as a sop for the aerospace industry. This was exactly the motivation that got President Nixon to finally approve the space shuttle in 1972. Since two-thirds of NASA's budget has historically gone to manned spaceflight, such a policy decision would mean a shot in the arm for this cold war artifact.

There is reason to believe that President Bush has already taken steps to move the country down just this path. John F. Ahearne, former Deputy Secretary of Defense, and Richard A. Stubbing, former Deputy Chief of the National Security Division of the Office of Management and Budget, argue that by May 1990 the Bush administration and the congress had already begun a shift toward "sizable increases in NASA research and development."[26] The plan, they believe, is to keep total R&D spending for defense and space constant. As defense declines further in the coming years, NASA will rise by the same amount in constant dollars. The net benefit will be stabilization of the aerospace industry in a volatile time. Seven of the top ten recipients of DoD contracts are also among the top ten NASA contractors — McDonnell Douglas, Martin Marietta, Lockheed, Boeing, Rockwell, TRW, and General Electric. The top ten defense contractors receive about 40% of DoD awards while the top ten NASA contractors receive about 60% of their funds. Ironically, then, the winding down of the cold war could feed manned spaceflight, just as the heating up of the Cold War fed it in the late 1950s and early 1960s.

The second path would move the United States in an entirely different direction. The Cold War rationale for manned spaceflight could collapse and this particular form of national behavior could be thrown back on its own merits. Congress and the public would ask for real return on investment, not just points in a prestige contest. Is manned spaceflight the most economical and productive way to conduct space activities? Is it a means to an end or an end in itself? By these standards, manned spaceflight would often be found wanting. It would certainly be exploited for some missions, such as rescue or repair of spacecraft in orbit. But sending astronauts into orbit for the sake of having them there would be seen as conspicuous consumption of a kind the U.S. can no longer afford.

Under this kind of scrutiny, the space station would be abandoned and replaced with a space platform, a permanent facility in orbit that could be visited, repaired, replenished, and even expanded by astronauts – but does not need to be permanently occupied. The shuttle, which is a spectacular but uneconomical and unreliable launch vehicle, would be reserved for those missions to which it is uniquely suited. Other missions would be launched on less expensive and more reliable launch vehicles, perhaps even Russian launch vehicles.

Which path the United States will choose cannot yet be discerned. The country is just now learning how to live without a Cold War, and for a whole generation of Americans that is entirely new and not altogether comfortable experience. However, frightening the Cold War may have been, it was at least the devil we knew. Now we face the more discomforting prospect of uncertainty – about a re-united Germany, an ambitious Japan, and an economically powerful Europe; many will cling to the Cold War as a less threatening enviroment. Those same people will continue to find manned spaceflight congenial.

Adding to the uncertainty is the equally unsure future of spaceflight in general. In spite of decades of predictions about the exponential rise in demand for space activity, demand seems actually to be leveling off, and it seems never to have come close to the optimistic predictions that were made for it.[27] Plans in the early 1960s called for 50 Apollo flights annually; having forgotten nothing and learned nothing, NASA predicted in the early 1970s that there would be 50 shuttle flights annually. One NASA official even said that the low cost of the shuttle would increase demand for spaceflight.[28] Of course, the opposite turned out to be true. The cost of spaceflight is so high that demand has remained low. All space activity save communications satellites still requires government subsidy, and the prospects for commercialization of space are if anything more remote now than twenty years ago when we knew less about the real costs of going into space. Since manned spaceflight is many times more expensive than unmanned, the prospects for this activity are bleaker than that for the overall picture.

The latter path, toward a more measured and rational space program, will have to contend with a nostalgic attachment to the Cold War and the space race it spawned. The choice between the paths is really a matter of timing. We may well have bases on the moon and Mars one day. We may have colonies at L-5 and elsewhere in the solar system. We may have tourism in earth orbit. But we won't do it soon. The Cold War, the only rationale for racing to achieve these goals, is over. Now we can do them in their time--when the technology is ripe, when the funding is available, when the need or the utility is manifest. That time is not yet.

Notes

1. Caldwell Johnson as quoted in H. S. F. Cooper, Jr., "Annals of Space, We Don't Have to Prove Ourselves," *New Yorker*, Sept. 2, 1991, p. 50.

2. Vernon Van Dyke, *Pride and Power: The Rationale of the Space Program* (Urbana, IL: University of Illinois Press, 1964).

3. The quotations are from Walter A. McDougall, . . .*the Heavens and the Earth: A Political History of the Space Age* (New York: Basic Books, 1985), pp. 141-45.

4. *Public Papers of the Presidents of the United States, Dwight D. Eisenhower, 1957* (Washington: GPO, 1958), pp. 720, 722, 728.

5. John Logsdon, *The Decision to Go to the Moon* (Chicago: University of Chicago Press, 1976), p. 20.

6. Logsdon, *Decision to Go to the Moon*, p. 35; McDougall, . . . *the Heavens and the Earth*, p. 225.

7. Inaugural Address, 20 Jan. 1961, *Public Papers of the Presidents of the United States, John F. Kennedy, 1961* (Washington: GPO, 1962), p. 1.

8. Logsdon, *Decision To Go to the Moon*, p. 111.

9. "Special Message to the Congress on Urgent National Needs," 25 May 1961, *Public Papers of the Presidents, Kennedy, 1961*, pp. 403-404.

10. McDougall, . . . *the Heavens and the Earth*, p. 291; Brian Harvey, *Race into Space: The Soviet Space Programme* (New York: Wiley, 1988), chaps. 5-6.

11. *Public Papers of the Presidents of the United States, John F. Kennedy, 1963* (Washington: GPO, 1963), p. 695.

12. Amatai Etzioni, *The Moon-doggle: Domestic and International Implications of the Space Race* (Garden City, NY: Doubleday, 1964).

13. W. Henry Lambright, *Presidential Management of Science and Technology: The Johnson Presidency* (Austin: University of Texas Press, 1985), pp. 104-107.

14. Edward Clinton Ezell and Linda Neuman Ezell, *The Partnership: A History of the Apollo-Soyuz Test Project*, NASA SP-4209 (Washington, NASA, 1978).

15. See especially William H. Chafe, *The Unfinished Journey: America since World War II* (New York: Oxford University Press, 1986); and Godfrey Hodgson, *America in Our Time* (Garden City, NY: Doubleday, 1976).

16. Charles D. Benson and William Barnaby Faherty, *Moonport: A History of Apollo Launch Facilities and Operations*, NASA SP-4204 (Washington: NASA, 1978), p. 529

17. The decision to cease production of the Saturn launch vehicle was announced in 1970, partly on the basis of the curtailment of the moon missions from nine to seven and the lack of firm plans for manned spaceflight after Skylab. *(Astronautics and Aeronautics, 1970*(Washington: NASA, 1972), pp. 10-12) By then, however, NASA was already promoting the shuttle, based in part on the argument that resupplying a space station with existing launch vehicles would be too expensive. The shuttle, of couse, has turned out to be more expensive than the Saturn.

18. Michael D. Lemonick, "Surging Ahead: The Soviets Overtake the U.S. as the No.1 Spacefaring Nation," *Time*, 130 (5 Oct. 1987):pp. 64-70; Nicholas Johnson, "1986: Very Good Year for Soviets; with *Mir* Leading the Way, Soviet Space Scientists Widen their Lead over their Rivals," *Space World* (Oct. 1987):p. 14; Ernest Volkman, "Meanwhile Back at the Space Race . . .," *Omni* 9 (May 1987):p. 40; *The Economist*, "Supremacy in Space," *World Press Review* 35 (Feb. 1988):p. 57.

19. Kurt Mendelsshon, "A Scientist Looks at the Pyramids," *American Scientist* 59 (March-April 1971):pp. 210-20.

20. Jean Gimpel, *The Cathedral Builders* (New York: Grove Press, 1983).

21. Thorstein Veblen, *The Theory of the Leisure Class: An Economic Study of Institutions* (New York: Modern Library, 1934), esp. chap. 4.

22. Ibid., p. 68.

23. Helen Codere, *Fighting with Property: A Study of Kwakiutl Potlatching and Warfare, 1792-1930*, Monographs of the American Ethnological Society, ed. by Marian W. Smith, XVIII (New York: J. J. Augustin, 1950); H. G. Barnett, *The Nature and Function of the Potlatch* (Eugene, OR: University of Oregon, Department of Anthropology, 1968).

24. In *The Rise and Fall of the Great Powers: Economic Change and Military Conflict from 1500 to 2000* (New York: Random House, 1987), Paul Kennedy has demonstrated the economic burden of great-power leadership. In a critique of this line of argumentation, Samuel P. Huntington has noted the role of investment and consumer spending. See his "The U.S.--Decline or Renewal?", *Foreign Affairs* 67 (Winter 1988/89):pp. 76-96.

25. "Soviets Suspend Manned Program," *Washington Post*, 12 April 1989, pp. A1, A18; Jeffrey M. Lenorovitz, "Soviet Space Program Reflects New policies Initiated by Gorbachev", *Aviation Week and Space Technology*, 131 (18 Dec. 1989):p. 52; Vera Rich, "Wings Clipped by Economics", *Nature*, 341 (21 Sept. 1989):p. 179; Peter M. Banks and Sally Ride, "Soviets in Space", *Scientific American*, 260 (Feb. 1989):pp. 32-40.

26. John F. Ahearne and Richard A. Stebbing, "Applying the Research & Development 'Peace Dividend.'" unpublished paper, [Durham, NC, May 1990]. Ahearne and Stebbing argue that as better policy would be direct support of those

technologies essential to our national welfare. Others agree. See, for example, Robert Kuttner, "Industry Needs a Better Incubator than the Pentagon," *Business Week* (27 April 1990):p. 16.

27. See Albert D. Wheelon, "Space Policy: How Technology, Economics and Public Policy Intersect," Working Paper Number 5, Program in Science, Technology, and Society, MIT, [1989], esp. chap. 2, "Getting into Space: Rockets and Shuttle."

28. George E. Mueller, "The New Future for Manned Spacecraft Developments," *Astronautics and Aeronautics*, (March 1969):pp. 24-32. "The space shuttle, by its very existence and economics, may generate the traffic it requires to make it economical." (p. 32) Mueller was predicting costs of $5 a pound to get payload into low earth orbit, or about $17 a pound in 1990 dollars. The actual cost today is about $8,000 a pound.

Part Two
CONTEXT

In this section six chapters address the context in which space policy is made and executed. They are arranged in order of scale, beginning with the chapter by Macauley on the overall Federal budget context. Next Telson describes the Congressional budget process. Read together, the two papers by Fellows and Guastaferro describe the incentives driving decision makers in government and industry and bring the scale of examination down to the program level. The final two papers of this part provide a transition to Part III (on specific programs) and indeed could fit into it. However, they seem to serve better as a conclusion to this discussion in that they show how context affects programs. Faget describes how manned programs have been delayed and increased in cost by slow execution. Coleman shows how the incentives in the system have driven science missions to become larger and larger. Faget argues that missions be done faster and Coleman argues for smaller ones. The two arguments are complementary.

Overall the picture is of a context which will be difficult even if managers are conscious of it and work to adapt to it. However, merely to complain about it is to court failure. Specifically, to plan programs that will take years of steady funding in an environment when funding is very uncertain is analogous to planning a launch vehicle without recognizing gravity. And as Faget shows, slowly executed programs become technologically obsolete before completion.

Due to limitations we could not explicitly cover the very important international dimension. Today the US is confronted not with one large rival in space but with several competitive allies. The rival is probably gone for the foreseeable future and the allies are healthier economically than the US.

Thus the real context for space programs and policy is even more difficult than pictured here, and must be recognized if programs and policies are to succeed.

Chapter 4
THE NASA BUDGET: FOR WHOM, FOR WHAT, AND HOW BIG?

Molly K. Macauley

THE AFFLUENT COUNTRY... IN MISUNDERSTANDING ITSELF IT WILL, IN ANY TIME OF DIFFICULTY, IMPLACABLY PRESCRIBE FOR ITSELF THE WRONG REMEDIES.
—GALBRAITH

Introduction

How big should the NASA budget be? In fiscal year 1990, the space agency's budget was about $12 billion. The 1991 Report of the Advisory Committee on the Future of the U.S. Space Program (the "Augustine Report") calls for NASA's budget during the 1990s to grow 10% per year after adjusting for inflation (in which case the agency's annual budget would be about $28 billion by 2000).[1] Are these the "right" amounts, in any sense of the word?

For those who believe without caveat in the effective working of a representative democracy, $12 billion, because it is the spending level that has been determined by the Congress and the executive branch, is indeed the "right" answer to the question "how much" to spend on public space activities today. After all, the level is the outcome of constituents' preferences that are publicly arbitrated through votes taken on the gamut of federal activities, from defense and entitlements to education, health, the environment, and science. It follows, then, that if ten percent future real growth is what the Congress (thus the voters) are willing to support, $28 billion in the year 2000 is also the "right" amount.

Yet the ideal process of voting arbitrage — that is, the logrolling and other vote trading mechanisms by which Congressional decisions are made — requires a setting characterized by full and certain information. But in fact, in the case of budgetary decisions regarding the level and scope of the nation's space program, incomplete and uncertain information tends to be the rule rather than the exception. The reason is at least twofold. First, decisionmakers' information is far from complete both about fundamental scientific aspects of space activity and about its costs and benefits (who knows what the magnetosphere is, let alone how worthwhile it is to understand it). This problem is true of much research and development activity, whether in medicine, high energy physics, or space, although the latter two examples are perhaps most challenging because their applications to everyday experience are less well-understood by the layperson. Second, there is a substantial amount of uncertain information (for instance, who knows how many shuttle trips and how many extravehicular hours a space station will require, let alone whether a return to the moon and a manned trip to Mars are possible for the presently estimated amount of $400 billion). While uncertain information plagues the conduct and management of research and development programs, the uncertainty of cost information in the case of space activities means that significant resources are at stake, as these activities tend to be among the most expensive federal R&D programs.

Even with better understanding of these information gaps, however, procedural aspects of voting in determining the space budget are far from ideal. Because of the way in which the Congressional budget process is organized,[2] the terms of exchange between a vote to increase the NASA budget and a vote to support some other federal activity are never clear. According to the process, dollars for nondefense discretionary spending (including space activities) first compete with dollars for entitlements (social security, Medicare, welfare) programs and defense activities. Then in a second round of the process, dollars for space compete with dollars for all other nondefense discretionary activities. Finally in a third round, space spending competes directly with specific nondefense discretionary activities including but not limited to housing and urban development, environmental regulation, science activities conducted by the National Science Foundation, and veterans' benefits. Thus, as Telson emphasizes, "the political dynamics change at every stage of the process." Supporters of NASA programs should in the first stage support increases in nondefense discretionary programs. Then, supporters who had been allied in the first stage compete against each other among subcommittees of the Appropriations Committee (the second stage). Finally, supporters of programs in a given subcommittee — who had been allies up to this stage — now compete against each other within that subcommittee (the third stage). Thus, the "currency" in which votes trade is far from transparent. A shuttle flight is the equivalent of how many jobs, houses, and veterans' benefits; how much national prestige; how many new recruits into the study of science and engineering. . . .

The Principal-Agent Problem

Thus the setting for effective public debate on the space budget is weakened from the start. To borrow from jargon, as in this case the jargon is a helpful analytical device, the difficulty is a "principal- agent" problem in which the principal (the public) desires the agent (NASA, by way of the Congress) to provide a successful space program. Key to the problem is that all of the parties have incomplete information about space activities.[3] For instance, the general public is not knowledgeable about the relative costs of the shuttle program and alternative space transportation, expendable launch vehicles (ELVs). The Congress has somewhat more information, but far less than NASA itself, particularly in terms of details about the different launch systems. The public and the Congress do, however, appear to be more excited when the shuttle flies successfully than when an ELV does (a shuttle launch often makes front-page headlines whereas an ELV launch usually does not). For this reason, the principal-agent relationship may rest not on delivery of a scientifically sound program (in which case most experts have suggested that NASA would rely more on ELVs) but on a media-centered one. Nothing in the principal-agent relationship operates to guarantee the right mix of shuttle and ELV flights. Commenting on space transportation issues more broadly, one policymaker has implicitly summed up the principal-agent difficulty: ". . . the current system requires NASA to pretend that its space goals are the nation's space goals. As a result, promises are made that cannot be kept and large expensive programs are started that have little chance of full and adequate funding."[4]

Why the relationship breaks down has several possible explanations. From the public's point of view, direct involvement in the space program, like voting in national and local elections, may simply not be worth it (as demonstrated by the perennial difficulty of space advocacy organizations in persuading the public to write to Congress). And indeed, passivity may be rational for at least two reasons. First, the public is generally well-removed from personal involvement in the program. This is true particularly in the program's scientific research components. Second, as Table 4.1 indicates, the share of the typical household's tax payment going to space is quite small — about 1% of the household's total federal tax payment, or about $52 each year. Thus, the principal tends not to be too demanding of the agent. For these reasons, the best space transportation mix for the nation might include just one successful shuttle flight a year to satisfy the public, and ELV flights to fulfill the rest of space transportation needs.

With only limited, perhaps "vicarious" public involvement, guardianship of the program passes to the Congress and to the executive branch. The problem for those who are responsible for national R&D on the public's behalf, however, is of course that while spending on space is a small fraction of the total tax dollar, it is a large fraction of the civil research and development dollar (about 20 percent — see the lower portion of Table 4.1. Competing for the dollar is a host of other activities. Panel (a) of Table 4.2 lists some of the major competition in the form of science initiatives proposed for the next decade or so. If national spending on non-defense research and development continues at the same rate as has prevailed since the mid 1970s, then the total annual expenditure for the projects in panel (a) *alone* (about $33 billion) will require a 50 percent increase in annual non-defense R&D funding (presently about $23 billion). Such R&D spending would increase the household's tax bill fairly significantly, by about 5%, or $300 per year. Any additional R&D projects would, of course, require an even larger increase in funding.

Other panels of Table 4.2 illustrate the broader environment in which the public guardians operate by listing some of the alternatives to spending on science. These alternatives represent the types of activities for which votes are traded with votes on space spending. As noted earlier, the trades arise because the Congressional bill that ultimately funds NASA is influenced by spending decisions made at other points in the Congressional budget process. Panel (b) represents

Table 4.1. The Allocation of the Annual Household Tax Burden, FY1990

All activities	Average Federal Tax Payment	
	%	$
National Defense	25	1300
Entitlements	44	2288
Interest on the debt	14	728
Health	5	260
Education	3	156
Agriculture	2	104
Environment	2	104
Transportation	2	104
Space	1	52
Other	2	104

Federal research and development activities (included in above totals)(basic and applied)

	All		Civil	
	%	$	%	$
Defense	65	203	-	-
Health	14	44	40	42
Energy	3	9	9	9
Space	8	25	22	23
National Science Fndtn	3	9	9	12
Agriculture	2	6	4	4
Other	5	16	17	18
	100	$312	101	$108

Notes and Sources: Amounts in percent are calculated as estimated 1990 federal budget outlays for each activity divided by total federal budget outlays using data from U. S. Office of Management & Budget, *Special Analyses: Budget of the U.S. Government 1990*, Tables A-5 and D-7. Amounts in dollars are calculated as amounts in percent multiplied by the average household federal tax payment using tax data from U. S. Bureau of the Census, *Statistical Abstract of the U.S.: 1990* (Washington, D.C., U.S. Government Printing Office), table 722. Tax data are for 1986.

amounts roughly commensurate with the NASA budget (about $12 billion in 1990). Panels (c) and (d) compare spending on scientific investigations in space with social programs of roughly equal size.[5]

Can the trustees of these funds – that is, the Congress – thus be expected to salvage the problematic principal-agent relationship existing between the public and NASA? Perhaps not. As is frequently the case with R&D programs, their job-creation potential can be more important to re-election than scientific output and other socialbenefits.[6] In particular, Congressional members are likely to maximize spending on jobs made available to their district (a localized benefit but a nationwide cost) rather than overall program success or nationwide social benefit. The tendency to treat jobs as benefits is particularly strong in the case of R&D, because there are few other tangible results that are appropriable by constituents. And, because the payoff to R&D activities is generally only in the long run, maximizing localized benefits despite their generalized cost is a rational response to the short-term re-election constraint facing the Congress. Thus, while voting by jobs may foster the principal-agent relationship at the local level, it does pose a conflict of interest with the public-at-large.[7]

The Principal-Agent Relationship:
The Science and Aerospace Communities

In addition to shirking by the public and re-election constraints of public decisionmakers, additional dimensions of the principal-agent problem involve the science community and the aerospace industry. The science community is at times a principal (when NASA is to act on its behalf) and at times an agent (when it is to deliver science to the public). The roles probably pose conflicting rather than consistent goals within the community. For instance, a 1988 member survey by the science honor society, Sigma Xi, showed that only 6% support the space station project. Yet many scientists also state that "big science" activities like the station become the "only games in town," thus the community feels forced to support the effort.[8]

The case of industry as agent is also problematic. Particularly in science and technology areas, industry typically operates in an environment fraught with technological and cost uncertainty. In this situation, if the job is not well done, the principal (the public) cannot know for sure if it is the agent's (NASA's) fault. This problem is

Table 4.2. Spending on Various National Science and Social Programs (billions of 1989 dollars)

(a) Proposed major national science activities and technology projects during next 15 years

Project	Estimated Total Cost	Estimated Annual Cost[a]
Superconducting super collider	$ 8.0	$ .5
Mapping human genome	3.0	.2
Space station	30.0	2.0
Manned mission to Mars	400.0	28.0
National aerospace plane	4.0	.3
Earth observing system	32.0	2.1
Total	$477.0	$33.1

(b) Selected social programs (FY 1989)

Project	Estimated Annual Costs
Elementary, secondary, and vocational education	$10.0
Higher education (financial assistance, student loans)	10.0
Social services (block grants, foster care, human development)	10.0
Housing assistance	10.0
Food and nutrition	21.0
Total	$61.0

(c) Space science research program (FY 1989)

Program	Estimated Annual Cost[b]
Physics and astronomy	.25
Life sciences	.05
Planetary exploration	.20
Solid earth observation	.02
Environmental observation	.13
Communications	.01
Total	$ .66

(d) Selected FY 1989 social programs with budgetary commitments commensurate with panel (c)

Project	Estimated Annual Cost
Summer youth employment	$ .7
Dislocated worker Assistance	.5
Job Corps	.7
Older Americans employment	.3
Low-rent public housing loans	.9

Notes: [a] Discounted present cost of project assuming four percent inflation and fifteen-year construction time.
[b] Adjusted from 1988 dollars to 1989 dollars using implicit price deflator for 1989.

Sources: Panel (a): NASA, Department of Energy, National Science Foundation. Panels (b) and (d): Budget of the U.S. Government, FY 1990. Panel (c): Congressional Budget Office, May 1988, The NASA Program in the 1990s and Beyond (Washington, D.C., CBO).

potentially quite large, of course, since typically close to 90 percent of NASA's budget is passed through to contractors.[9] How the risk of cost overruns is distributed between NASA and contractors, and how accountability for success or failure is perceived in the public and the Congress, are key elements of the principal agent problem.

Strategic Behavior: Implications of the Principal-Agent Problem for the NASA Budget

The principal-agent problem as described above leads directly to predictable behavior on the part of the players (NASA, the Congress, and the science and aerospace communities) in arguing the size and allocation of the NASA budget. As long as the budget remains a relatively inconsequential share of the taxpayer's burden and as long as some visible success (like a successful shuttle launch now and then) is apparent, the public remains on the outside of the budget game.

The challenge for referees of the game — say, the Office of Management and Budget or the House and Senate Budget Committees — is that the difficulty of sizing the NASA budget and allocating it across various activities is largely related to the difficulty of estimating nonjob program costs and benefits. Industry and NASA experts are in the best position to estimate costs (even though the estimates are subject to uncertainty), and referees lack the time and resources to go over cost estimates in detail. The consequence is that NASA has incentives to package its budget proposal as an all-or-nothing approach — for example, fund the station or the Earth Observing System at the level NASA requests, because it can't be carried out otherwise.[10] It appears that the all-or-nothing approach generally costs more than the Congress is willing to spend, because the Congress typically replies with a smaller-than-requested NASA budget. This course of events is repeated annually; thus, because the Congress typically responds with a smaller budget, the incentive is there for NASA to enter the negotiations by proposing high.[11] If a worthwhile space program is nonetheless forthcoming, even if reduced in scope or efficacy, then this annual ritual may be relatively harmless. The opportunity is there, however, to use the smaller actual budget as an excuse for poor performance, and it is difficult to ascertain if the budget was in fact adequate for the task.

In addition, the parties to the budget negotiation have the opportunity to selectively disclose information. Examples are understating total project costs (if too large, the amount may be deemed ludicrous by the Congress), or, in claiming that a project's scope is all or nothing, exaggerating the importance of what the agency is giving up and minimizing the importance of what is got in return, rather than offering a menu of alternative approaches. The information disclosure problem may be real, in that costs are difficult to estimate; it may also be made worse if NASA contractors withhold information or ideas about alternative approaches. In the extensive literature on these types of bargaining situations[12] selective disclosure is not considered dishonest; rather, it reflects strategic behavior induced by the negotiating situation.

Why the ritual of the large, all-or-nothing budget persists is partly related to incentives inherent in the political process and partly related to the difficulty of specifying and measuring output or success in the nation's space activities. Unlike obtaining an AIDS cure (an output metric for biomedical research funds) or a high graduation rate among high school students (an often-used benchmark for the efficacy of education funding), what constitutes a job well done in the case of space? As discussed earlier, to the Congress output is largely measured as jobs. To other parties, typical answers are (A) a successful shuttle launch per se; (B) publishable findings from a microgravity experiment (perhaps enabled by a successful shuttle flight); and (C) a profit-making new pharmaceutical (perhaps enabled by the microgravity experiment, in turn made possible by the shuttle flight). Answering "all of the above" is not allowed, as so doing is useless as a guide to resource allocation. To see this, note the alternatives by which (B) might be brought about — by funding more sounding rocket flights, funding recoverable payloads launched by ELVs, or buying research time on the Soviet space station. Goal (C) might be brought about through increased funding on ground-based pharmaceuticals R&D. In these cases, neither (B) nor (C) requires (A), the shuttle. Choosing one of the responses, then, would clarify the steps needed to reach any of the goals. In turn, scarce budget dollars could be better allocated.

As noted earlier in discussing Table 4.2, the broader context of all of the federal programs to which scarce dollars might be directed is also important in considering what success is in the case of space, and how the nation chooses to spend money to obtain it. The relative social impact of an AIDS cure or graduation rates are more readily estimable than the value of success in space. In other words, an approximation of AIDS exposure and death rates can be made, together with a rough estimate of the social impact in terms of loss of life. To be sure, controversy has always and will always surround dollar calculations of a human life, not just the calculation but whether doing the calculation is ethical (can you put a value on life). But such estimates are routinely done in the private sector (say, for insurance) as well as by government (in health, safety, and environmental regulation). Similarly in the case of graduation rates, the estimated productivity gain from increments of education could be used to measure the benefit of education spending. The point is that these estimates provide a rough benchmark of how much taxpayers' money to spend in researching AIDS or funding education. Presumably the nation would not support allocating 100% of the federal budget to either one; just how much to allocate can at least have as an upper bound the maximum estimated gain of the activity.

The case of space is quite different. Except in the case of satellite communications, and, perhaps, the Apollo program, benchmarks so far have been lacking as to the value of space activities. It is outside the scope of this paper, but not outside the scope of methodological advances in the social sciences, to consider how to conceive of and then measure some of these benchmarks. (The next section of the paper discusses how these benchmarks might be estimated.)

The implications of this discussion for sizing the NASA budget are twofold: given some consensus as to what constitutes success, budget dollars could be guided by the value placed on the activity compared with other activities. If the nation wants to increase science and engineering students, then the most effective approach may not be by way of space. If the goal is jobs or national prestige, then it may not be by way of microgravity research undertaken on sounding rockets. If science is the aim, then it may not be best obtained by way of the space station.

Rectifying the Principal-Agent Problem

Given a passive public, the re-election constraint, substantial uncertainty in the technology, and output or success that is difficult to specify and measure, it is hard to know whether increasing costs and program setbacks in space activities are the consequences of misinformation, poor management, poor quality engineering, or bad luck. Indeed, whether the NASA budget is the "right" size depends on a flawless principal-agent relationship then, but the above discussion suggests that flaws prevail. If so, what steps might correct the imperfections? Some possibilities are:

(i) Can fuller information and accountability on behalf of various parties — the public, NASA, the science community, industry, the Congress — be brought to bear on financing and operating the space program?
(ii) Can greater rigor and, perhaps, increased certainty be brought to estimates of benefits, costs, or both?
(iii) Can an improved system of checks and balances be implemented, to structure the agents' incentives in such a way that the principal's expected net gain is as high as possible?

The remainder of this paper considers the first and part of the second questions (benefit or output measurement) in some detail under the theme "elevating the quality of debate." Extensive discussion of the other part of the second question, improved cost estimation, is outside the scope of the paper largely because the topic pertains to a sizeable and growing literature on the theory of contracts. While NASA procurement reports indicate that the bulk of contracts are incentive-based, their specific structure and whether they exploit any useful modern-day developments in the literature are topics for future consideration. Guastaferro (in this volume) takes a large step in this direction.[13]

Discussion of the third question is also deferred although there has been no shortage of suggestions for institutional change, as illustrated by the following sample of proposals. They are arranged roughly in decreasing order of the extent of institutional reform necessary to implement them:

— announce a prize of $400 billion for the first entity or consortium to send a successful manned mission to Mars. If there are no takers, then the project may well be technically infeasible at that level of willingness to pay;[14]
— establish a Space Frontier Agency (to conduct operational activities like operating the shuttle);[15]
— develop a quasi-public spaceport or launch corporation, supervised by the Department of Transportation, to manage launch ranges and tracking and data;[16]

- reorganize NASA centers as directorates, responsible for "core functions" such as aeronautics R&D, space science, exploration, etc.;[17]
- reorganize NASA centers as government-owned, contractor-operated laboratories (GOCOs) like the Jet Propulsion Laboratory;[18]
- experiment with a pilot program that gives grants (or "vouchers") to space scientists to permit them to choose the design, financing, and timing of space transportation that best accommodates scientific aims and scientists' preferences for payload design;[19]
- experiment with a pilot program, modeled on a newly established NSF program, to fund by way of a rapid, abbreviated review process small-scale innovation in space science methodology or research (the science counterpart to small business innovation research).

The general emphasis of each of these possible innovations is to shore-up the principal-agent relationship by imposing incentives to elicit a different type of strategic behavior among the players. The "prize" for a Mars mission constitutes a direct economic incentive to develop workable technology that costs no more than what the nation is willing to pay for the mission. Vouchers and rapid funding of small-scale innovation are examples of indirect economic incentives that might sharpen opportunities to focus on space science rather than job creation as an objective of the nation's space activities.[20] Reorganizational schemes such as the Space Frontier Agency, launch corporation, and redesigning NASA centers alter the nature of responsibility and accountability among NASA, the aerospace community, and the Congress.

Elevating the Quality of Public Debate

Aside from contract reform and institutional change, are there "nearer-term" actions to improve policy? The following caricature, by paraphrasing some of the present terms of debate in the space program, highlights missing and mis-information. So doing may suggest where chinks might be filled to begin to shore up the principal-agent relationships:

The Debate Caricatured

A member of Congress: "The space program is vital to the nation's economy. It gainfully employs our most productive resources, inspires our young people, and keeps the nation competitive."

The public: "The shuttle proves that we have what it takes; we've got people in this country who have the 'right stuff.' Oh, and I remember the pictures from Voyager or whatever it was were pretty neat. Mir? What did you say that is?"

A decisionmaker in the executive branch: "We need significantly improved scientific understanding of climate change before we take actions that could cost the global economy upwards of several hundred billion dollars. A space-based earth observing system demonstrates our commitment to this problem."

Another decisionmaker in the executive branch: "We need to go in swinging on the space budget. Every dollar we spend on space is a dollar not spent on entitlements programs that don't work anyway."

An astronomer: "Probably 90 percent of the usefulness of the Hubble (Space Telescope) is compromised by the mirror's spherical aberration."

Another astronomer: "Probably 40 percent of the usefulness of the Hubble is compromised by the mirror's spherical aberration."

A scientist not working in a space-related field: "I've heard the NASA presentations on research opportunities in space, and I tell you, they are a joke. Space scientists sell out to NASA all the time."

An industry leader: "The space program is vital to the nation's strength. It gainfully employs our most productive resources, inspires our young people, and keeps the nation competitive."

A NASA deputy: "The youthfulness of the nation's merely thirty-year-old space program belies its tremendous contribution. That contribution, of course, extends beyond the significance of a shuttle flight to pervade virtually every aspect of daily life. The fruits of the space program are right in the home in the convenient packaging of breakfast drinks and the light weight, but high strength fasteners in autos, on garments, and in the workplace. Perhaps more importantly, space has brought home the highest technology, like making computers portable enough for the lap-top. Even more importantly, space has saved lives -- witness the miniaturization of pacemakers and the development of materials for prosthetic devices."

The usefulness of a caricature is that it sets in high relief the "focal" points around which parties to discussion frame their arguments and which figure prominently but often implicitly in public debate. The present dialog demonstrates several flaws, including those listed below.

Flaws in the Dialog
— *The lack of a basic level of informed understanding.* Numerous studies have shown that science skills are weak among the population as a whole. Mir is virtually unheard of; what Voyager photographed and why it is significant are generally forgotten.
— *Erroneous key assumptions.* Job creation appears to drive much of the Congressional will in supporting space activities, but jobs are a cost of the program, not a benefit. The garnering of national prestige or the demonstration of national technological prowess is also an oft-cited justification for space activities. Yet the pursuit of space activity is an expensive route to these options, and if not carefully balanced, their pursuit can distort other aspects of the program. The relegating of science missions to shuttle transportation is an example; the shuttle is a highly visible manifestation of some technical skill but it is not always the best transportation mode for science missions. In addition, because space activities are also a highly visible route to international glory, failure, as well as success, is noticed. Nor are space activities the only route to prestige or technological advance.
— *Tenuous key assumptions.* The bulk of jobs generated in space activities are probably "medium-tech" not "high-tech," in the sense of consisting largely of assembly-line jobs in aerospace. "Spillovers" like TANG, velcro, or miniaturization are also a prevalent rationale for space projects, but they are products whose development, if deemed in the national interest, could have been directly funded, most likely at significantly lower cost. The value of space-gathered

information in decisionmaking — perhaps most notably promised in the proposed mission to planet earth (intended to gather observations on global environmental change) — depends critically on timeliness and accessibility of the information. And, the value of the information is highly dependent on whether the right hypotheses have been formulated (do we know what to look for in the data?) and whether models to incorporate the data are good enough.[21]
— *Contradictory key assumptions.* We may compromise scientific integrity if we justify public works on the basis of scientific merit. Large-scale, high-public-visibility efforts, like the shuttle program, and a smaller scale, highly focused scientific effort like the Cosmic Background Explorer may not be commensurate in their power to attract and retain young people in the study of science and engineering. If true, then how is a balanced research and transportation program structured?

Given these observations, can the distortions in the caricature be usefully refocussed — that is, can the debate be made more constructive? Table 4.3 is an attempt to do so. It is a highly stylized juxtaposition of principal and agent perspectives of benefits (or objectives) and costs of the space program. The table includes direct benefits, represented as actual tangible impacts on society. Spillovers do not count as direct benefits; something which might be called improved decisionmaking does count (easy to see in the case of space-based weather information). Some benefits are less tangible but are nonetheless a direct outcome of space activity, such as expanded human knowledge from space science or the generation of national pride or geopolitical advantage. Benefits that are indirect — arising from a side-effect, rather than a direct aim — include spillovers or the maintenance of certain technological industries.

Clearly there is room for argument in how entries are categorized in the table; for example, some might claim that maintaining technological infrastructure is the direct objective of space activity. The point is that such argument can be accommodated by the framework of the table, and that by making such argument explicit, the nation can better reach its objectives. So, if maintaining industry is the aim, the relevant question is whether so doing by way of the space program is the best way, or might there be alternatives (direct funding, tax credits, etc.).

Table 4.3. Stylized Illustration of Organizing Principles for Considering Societal Benefits (Objectives) and Costs* of U.S. Space Program (or, possibly, a specific space activity within the program)

		Benefits (Objectives)	Costs
Direct	Tangible	Actions taken (or not taken) -- e.g., improved decisionmaking in light of information generated	Cost of spacecraft and associated hardware, launch, and other facilities Salaries, wages, and costs of management, administration, and other overhead
	Less tangible	Expanded human knowledge Generation of national pride or geopolitical advantage Thrill of research and exploration	Environmental degradation from space activities (e.g., space debris, launch and testing site pollution)
Indirect	Tangible	Generation of technical progress Maintenance of technological industries Attraction of students to science and engineering Improved public education and enhanced science and world awareness	Diversion of physical and human resources from other science, industry, and public programs
	Less tangible	Gain in world prestige (if successful)	Loss in world prestige (if failure)

* The table is merely illustrative and not intended to be comprehensive.

Source: Adapted from generic framework in Public Finance in Theory and Practice, Richard A. and Peggy B. Musgrave (New York, McGraw-Hill), 1989, Table 9-2.

Table 4.3 also lists costs that are direct and indirect, and tangible and less tangible. Of special importance on the cost side is the notion of opportunity cost − using space as a means of generating national pride comes at a cost of resources foregone. If the organizingprinciples underlying Table 4.3 are brought to the fore in public debate, then maybe objectives could be pursued more effectively, wasting as little as possible and getting as much for the effort as we can.

The table is not meant to be comprehensive; moreover, there may need to be a series of such tables each designed for a specific activity (say, earth observations, planetary exploration) and only later organized for the space program as a whole. But the idea of the table is intended to suggest the need to take the following actions in public debate:

1. distinguish costs (jobs) from benefits;
2. make assumptions explicit;
3. openly distinguish and publicly arbitrate the relative weighting of means and ends;
4. acknowledge the irrelevance of sunk costs; and
5. investigate methodologies for metrics to monitor program efficacy − that is, for imputing value to the benefits of space activity.

The first recommendation, distinguish costs and benefits, relates to concerns such as those discussed earlier which confuse job creation as a benefit when in fact it is a cost. The second recommendation, make assumptions explicit, seeks to eliminate or reconcile contradictory assumptions and ensure that no assertions disguised as objectives or costs are overlooked. By way of example, does international cooperation to share space hardware and thereby reduce the costs of a project (on the right hand side of Table 4.3) subsume an individual country's chance to demonstrate international preeminence (on the left hand side)? Such tension has already surrounded international collaboration on the space station and in space transportation. How does the scale of the activity (the right hand side) correspond to the number of opportunities to attract and retain students (the left hand side)? For instance, does a large scale effort like the proposed earth observing system offer more, or fewer, graduate student opportunities, say, than a smaller scale effort? And, how are scale, public visibility, and graduate training related? As another example, the nation has never debated a space budget without a manned component. What happens if we separate manned and unmanned activity in the budget process — how big would each budget be, and is support for one independent of support for the other?

The third suggestion, distinguish means and ends, pertains to issues such as the net benefit derived from successful spacecraft or hardware construction and flight (the putative means), versus the generation of unique and useful scientific or engineering data (the putative ends). What is desirable here is that a balance be struck so as not to unduly compromise the pursuit of either activity, if both are determined to be national objectives (that is, if both are ends). In the event that one is determined to be a means to attaining the other, the conduct of policy might be a little easier. The Augustine report (pg. 5 and elsewhere) urges that science be the "fulcrum" of the entire civil space effort, stating that "the space science program warrants highest priority for funding. It, in our judgment, ranks above space stations, aerospace planes, manned missions to the planets, and many other major pursuits which often receive greater visibility." Exactly what constitutes science is not clear (many might claim that propulsion technology for the aerospace plane or manned planetary missions are science) especially as, despite the above statement, the report goes on to identify manned missions as one of two focuses (the other is monitoring global climate change).

Suggestion four, acknowledge the economic irrelevance of sunk costs, is perhaps the hardest to implement. If after substantial spending on a given program the expected benefits are unapparent or significantly reduced, then the program should probably be terminated. Yet examples of perpetuation abound. The Clinch River Breeder Reactor, the centrifuge uranium enrichment plant, and some projects of the Synthetic Fuels Corporation are big-ticket energy R&D programs extended beyond their useful scientific lives by principal-agent problems. Prominent in these problems are the huge political sunk costs which keep the programs alive.[22] An additional point is well expressed by Ahearne, who distinguishes the potential relevance to science of the experience of learning from failure (the scientific relevance of sunk costs), but the irrelevance of past expenditure in the decision to continue a project when it has clearly failed (the economic irrelevance of sunk costs):

> Proving a hypothesis wrong or a promising approach to be a blind alley is not a failure in basic research f both results advance knowledge. Finding a technology uneconomic or infeasible may not represent a misuse of federal funds. The research and early engineering development stages are for exploration of new

concepts and for preliminary determination of economic and technological feasibility. If work at these stages is done well, some proposed programs will be canceled. These cancellations will represent appropriate uses of federal funds. Failure includes funding poor experimental design, mediocre work, and publication for the sake of publication. Imprudent use of federal funds occurs when the early stages are not implemented carefully or examined rigorously, and projects advance that should not.

The structure brought by Table 4.3 could serve as a scorecard throughout the program lifetime.

Item five attempts to translate components of Table 4.3 to the world of measurement, although it need not take place in dollars. If space policymakers took such a step, outlining for each program what constitutes success, they would merely be coming up to speed in the world of public policy analysis and programmatic defense during budget negotiations. Much of the rest of the intangible world has begun to explore ways to impute value to things that are hard to measure, such as environmental quality, habitat sustainability, or the amenity value (as opposed to the easier-to-value health benefits) of clean air and water.

How might value be imputed to the benefits of space activities? In the case of actions taken or not taken (see entry in Table 4.3 and earlier discussion of that entry), the procedure is conceptually straightforward. An example is the extent to which satellite-derived earth observations data can improve decision making concerning the desired level or rate of reductions in greenhouse gas emissions, or monitor compliance with international or national environmental protocols. Benchmark information is available on the costs of reducing greenhouse gas emissions.[23] Accordingly, if satellite data permitted us not to have to take these actions, or to undertake them at a more gradual rate, then the value of the data would be measured by the cost of the actions not taken or the savings in economic dislocation no longer made necessary if actions are phased in.

Less tangible is the value of knowledge, although metrics used in the scientific community (bibliographic citations, peer review) are possible. The issue here is how such metrics might be incorporated and weighted in public debate, say, in balancing various activities. Also less tangible is the benefit ascribed to national prestige or geopolitical advantage. Yet for a given activity, if we have an estimate of the direct costs (the upper right-hand side) and tangible benefits (the upper left-hand side), then a plausible inference is that prestige or geopolitical advantage must be worth at least any excess of direct costs or we would not carry out the activity. That is to say, if the geopolitical advantage were not worth at least the excess costs, then a lower-cost alternative means of international influence should be pursued.[24]

Concluding Comments

How exactly might the stylized picture in Table 4.3 improve the principal-agent problem? It is not necessarily the case that such a figure needs *actually* to be constructed for inter- and intra-programmatic deliberations. However, it might be useful to do so, because despite all its limitations, the framework is a helpful way to summarize argument and ascertain if all pertinent information has been brought to bear. It also forces decisionmakers to state assumptions explicitly so that the *reasons* for ultimate decisions are clear. In fact, any other process is implicitly just such an analysis without assumptions explicitly stated and perhaps without full information. Thus, some decision framework like Table 4.3 might usefully be required of all NASA programs at their outset, and used as a benchmark to measure progress throughout program execution.

Seeking to elevate debate in such a manner also makes clear that part of the burden rests with heretofore quiet players, including the public and the space science community. While one percent of the typical household's tax bill may render space presently unimportant enough to worry about, the space station, Earth Observing System, and human exploration initiatives would dig much deeper into pockets. Moreover, a public that appears increasingly to care about environmental amenities, global sustainability, and math and science skills may indeed value space research for knowledge's sake. Then again, it may not. Finding out is important.[25]

Similarly, the science community could play a large role in raising the quality of the debate insofar as it can enunciate the scientific benefit of knowledge gained from space. Reports that tend to be circulated only within the community suggest that these benefits are indeed large (having radically changed our understanding of the planet and the universe).[26] If so, such benefits should expressly be included in debate. The responsibility, expertise, and as budgets tighten, the incentives to do so rest with the community.

Notes

1. See the "Report of the Advisory Committee on the Future of the U.S. Space Program" (the Augustine Report) (Washington, D.C.: USGPO, December 1990), p. 4.

2. M. Telson, "NASA and The Budget Process," Chapter 5, this volume.

3. The principal-agent problem has a long history, particularly in the fields of finance (where the construct is used to describe the relationship between a corporation's shareholders and managers); insurance (where the principal and the agent are the insured and the insurer, respectively); and medical care (where the problem illustrates the behavior of the physician vis a vis the patient). Key to each case is the fact that the principal and agent differ in the amount and type of information each has. Their behavior is affected accordingly. For discussion and other applications, see S. Ross, "The Economic Theory of Agency: The Principal's Problem," *American Economic Review* 63, (1963) pp. 134-139; M. Harris and A. Raviv, "Some Results on Incentive Contracts with Applications to Education and Unemployment, Health Insurance, and Law Enforcement," *American Economic Review* 68, 1978, pp. 20-30; S. Shavell "Risk Sharing and Incentives in the Principal and Agent Relationship," *Bell Journal of Economics* 10, 1979, pp. 55-73 and K. Arrow, "The Economics of Agency," in J. Pratt and R. Zeckhauser, eds.,) *Principal and Agents: The Structure of Business* (Cambridge, MA, Harvard Business School Press, 1985.

4. R. DalBello, "Space Transportation and the Future of the U.S.Space Program," in R. Byerly, ed., *Space Policy Reconsidered* Boulder, CO, Westview, 1989, p. 80.

5. Many of the activities in panels (b) and (d) are under the aegis of the Committee that funds NASA, thus they compete directly with dollars for NASA.

6. For extensive discussion and other examples of the political economy of technology programs, see L. Cohen and R. Noll, *The Technology Pork Barrel* (Washington, D.C., Brookings Institution, 1991.)

7. How generalized is the cost and localized is the benefit of NASA spending are suggested by the following observations. If NASA expenditures are in fact "public works," then the nation benefits most when NASA spending boosts employment in areas where unemployment rates are high. Yet NASA spending tends to do the opposite. Of the 22 states with unemployment rates higher than the national average during 1980 to 1988, 15 states received below-average NASA spending in 1989 (on a per-state basis). Of the top ten states receiving the highest amounts of NASA spending, 6 states have unemployment rates less than the national average. If states with NASA centers are omitted from the calculation (because spending in these states is well above the average f NASA geographic expenditures are a highly skewed distribution) then even in this case, over 70% of states receiving an above average share of the space budget have unemployment less than the national average. Thus employment enhancing efforts of the pork barrel are far from clear. Note also that the re-election constraint may not be fully operative in the short run, given an apparent "lock-in" effect of NASA centers. That is to say, states with limited Congressional representation on key committees receive large percentages of NASA funding if they are home to a NASA center. Outside of states with NASA centers, however, NASA spending by state correlates well, although not perfectly, with the size of the Congressional delegation. The noise in the data probably reflect currency differences alluded to earlier f we need to know the exchange rates of, say, veterans' benefits vs. NASA spending to track the intricacies of voting results. (Unemployment rates are in

Statistical Abstract of the U.S. 1986 (table 690) and *1990* (table 629); NASA spending by state is in NASA, *Annual Procurement Report FY 1989*.)

8. The survey indicated skepticism toward other so-called "big science" projects as well, with only 4% and 2%, respectively, supporting the human genome project and the superconducting super collider. The surveyed scientists were asked to choose the "three best uses of public funds for scientific research." Top scorers were untargeted individual research awards, research on biosphere/geosphere systems, and research on AIDS. Although the survey sample does not fully represent all scientists and engineers, the respondents represented more than 160 disciplines from academia, industry, and government and was deemed "representative" of the community. See Sigma Xi, *Sketches of the American Scientist* (Research Triangle Park, NC, Sigma Xi, September 1988) and summary review in "Poll Draws Portrait of U.S. Scientists' Views," *Chemical and Engineering News*, 7 November 1988.

9. In FY 1989, the most recent year for which data are available, 88 percent of NASA's budget was allocated to procurement. Of this, 79 percent went to industry, 6 percent to educational and nonprofit institutions, and 5 percent to government agencies, and 10 percent to the government-owned, contractor-operated Jet Propulsion Laboratory.

10. See W. Rogerson "Quality vs. Quantity in Military Procurement," *American Economic Review*, 80, 1990, pp. 83-92, for discussion of this strategy in the case of Department of Defense procurement. He focuses on political economy results in the choice between quality and quantity in defense procurement.

11. The tendency is pervasive throughout agencies; in fact, in recent history there are very few instances where an agency proposed a budget cutback. The most recent example is the Department of Education in the early 1980s.

12. See, for example, H. Raiffa *The Art and Science of Negotiation*, Cambridge, MA, Harvard University Press, 1982.

13. See also J. Quirk, M. Olson, H. Habib-agahi, and G. Fox, "Uncertainty and Leontief Systems: An Application to the Selection of Space Station System Designs," JPL Paper No. 28 (Pasadena, CA, Jet Propulsion Laboratory, 1987) for a theory of the nature of cost overruns in large, complex, uncertain systems. They suggest that program engineers have in mind a probability distribution over which actual costs will range but typically report the lower of these distributions. Quirk and coauthors apply their model to the space station. If this literature is correct, then policymakers need a mechanism and interpretive skills to elicit the parameters of the distribution during budgetary deliberations. Such a recommendation is remarkably similar to recommendations now emerging in the environmental, health, and safety community, where decisionmakers are being urged to consider ranges of estimates and their associated confidence intervals, rather than point estimates (for example, see A. Finkel, *Confronting Uncertainty in Risk Management* (Washington, D.C., Resources for the Future, 1990)).

14. The prize is modelled after Gump's suggestion (D.P. Gump, "Put Space Programs in the Capitalist Orbit," *The Wall Street Journal*, 18 July 1990) and has numerous precedents. Charles Lindbergh's Atlantic crossing reportedly was inspired at least in part by a $25,000 prize posted by a hotelier; successful development of an astronomical clock key to maritime travel was also spurred by a financial prize. Economic historian Joel Mokyr comments that "Wherever these kinds of prizes have been proposed, they've almost always make a big difference . . . (and) can be a fantastic idea to spur innovation." (See reference in *Econ Update*, vol. 5, no. 1,

September 1990, p. 4-5, to J. Mokyr, *The Lever of Riches: Technological Creativity and Economic Progress*, (New York, Oxford University Press, 1990)).

15. For discussion, see J. Bennett and P. Salin "America's Space Crisis: The Case for Institutional Reform," mimeo, June 14, 1986).

16. See Bennett and Salin.

17. See discussion in Committee on Space Policy, National Academy of Sciences, National Academy of Engineering, *Toward a New Era in Space* (Washington, D.C., National Academy Press, 1988) and in the Augustine report.

18. R. Giacconi, "Science and Technology Policy: Space Science Strategies for the 1990's," in R. Byerly, ed. (*Space Policy Reconsidered*, Boulder, CO, Westview, 1989, pp. 83-104) discusses advantages of the GOCO framework; see also the Augustine report.

19. M. Macauley and M. Toman, "Providing Earth Observation Data from Space: Economics and Institutions." *American Economic Review*, May 1989, pp. 38-41, outlines the voucher idea.

20. The Augustine report argues that science become the premier objective of the nation's space program, but fails to suggest explicit mechanisms to realign the principal-agent relationship to make it consistent with this objective.

21. See M. Macauley and M. Toman, "Providing Earth Observation Data from Space: Economics and Institutions,: *American Economic Review*, May 1991, pp. 38-41 for discussion of the proposed Earth Observing System.

22. These examples are from J. Ahearne, "Why Federal Research and Development Fails," Resources for the Future Discussion Paper EM88-02 (Washington, DC, Resources for the Future, 1988). See also L. Cohen and R. Noll, *The Technology Pork Barrel* (Washington, DC, Brookings Institution, 1991) for reasons why some of the projects lasted, and R. Byerly and R. Brunner, "Future Directions in Space Policy Research," in R. Byerly, ed., *Space Policy Reconsidered* (Boulder, CO, Westview Press 1989) for discussion of political costs ("chips") in space activities.

23. For instance, see W. Nordhaus, "Count Before You Leap," *The Economist*, July 7, 1990.

24. By way of example, consider the series of shuttle flights during the three full years of shuttle operation, 1983 to 1985. These flights launched over a dozen communications satellites at below-cost fees. An interpretation of such a practice of "promotional pricing" is to showcase the shuttle. In this case, demonstrating prowess cost about $1.5 billion. This is the difference between (a) the real cost of the shuttle flights and (b) the alternate cost of using unmanned launch vehicles (when possible) for a given payload. By way of comparison, the difference between (a) and (b) is equal to about 10% of 1985 spending on international affairs. (It is true that revenue from launch fees reduced but did not fully offset this deficit. But below-cost fees induced some customers to use the shuttle who otherwise might not have, had the shuttle been priced closer to real resource cost. Thus, below-cost fees distorted decisions throughout the space community. Nor is it clear that the subsidies were later offset by scale effects or learning by doing, because in fact, the cost of space transportation operations has increased, not decreased, in real terms.) Thus, if shuttle flights brought international prestige benefits greater than commensurate levels of spending on other international activities, then shuttle flights are a cost-effective way to buy prestige. But another question must also be asked: whether just one successful shuttle flight a year would bring about the same amount of prestige. If so, then NASA resources could be reallocated from shuttle flights to other space activities.

25. These types of activities tend to be highly correlated with income, thus present prosperity boosts interest in them. A recession in the domestic or world economy may significantly lessen willingness to pay.

26. For example, see the stocktaking and agenda-setting in National Research Council, *Space Science in the 21st Century* (Washington, D.C., National Academy Press, 1988) and National Research Council, *1990 Update to Strategy for the Exploration of the Inner Planets* (Washington, D.C, National Academy Press, 1990). To be sure, these reports are publicly available, but has their message been summarized articulately for the Congress and the public?

Chapter 5
NASA AND THE BUDGET PROCESS

Michael L. Telson[1]

NO MONEY SHALL BE DRAWN FROM THE TREASURY BUT IN CONSEQUENCE OF APPROPRIATIONS MADE BY LAW.
—U. S. CONSTITUTION

Introduction

This chapter describes the context in which NASA budgets are made and discusses some of the issues that have been raised since Gramm-Rudman-Hollings (the Balanced Budget and Emergency Deficit Control Act) was enacted in late 1985. While there is broad support for space programs, funding for them must go through a rigorous process. The ultimate budget-making reality in this environment is the zero-sum nature of the budget process. The implications of this are explored here.

The NASA "Core Program"

Budget making for NASA since 1986 has been particularly difficult because of the collision between the resources needed to implement NASA's so called "core program",[2] and the limited availability of "nondefense discretionary" budgetary resources over the past decade.[3]

The Congressional Budget Office of the U.S. Congress (CBO) published a study in 1988 that showed the dimensions of the NASA core program.[4] It involves, roughly, growing the agency budget from $12.3 billion in fiscal year 1990 to $16.4 billion by fiscal year 1999

Table 5.1: Total NASA Budget Projected in Various New Initiative Studies (in billions of 1989 dollars).

Fiscal Year	1988	1989	1990	1991	1992	1993	1994	1995	1996	1997	1998	1999	2000
Baseline	9	9	9	9	9	9	9	9	9	9	9	9	9
"Core Program"	9	10.9	12.3	13.2	13.8	14.4	14.4	14.4	14.2	14.1	15.4	16.4	16.4
AIAA	9	12.8	14.8	16.5	17	18.7	19.5	19.7	20	20.2	20.2	21	22.3
National Commission on Space	9	9.7	11.6	12.6	13.2	14.7	16.3	17.9	18.9	21.1	22.1	23.2	24.2
Ride Report Moon and Earth	9	13.65	15.05	15.95	16.55	25.7	25.7	25.7	25.5	25.4	26.7	31.7	31.7
Ride Report Mars and Earth	9	16.65	18.05	18.95	19.55	28.5	28.5	28.5	28.3	28.2	29.5	32.7	32.7

Source: Congressional Budget Office

Table 5.2
NASA Historical Budget Totals and President Bush Projections
(in millions of dollars)

Fiscal Year	1985	1986	1987	1988	1989	1990	1991	1992	1993	1994	1995
Budget authority	7,573	7,807	10,923	9,062	10,969	12,324	15,240	17,638	19,258	20,261	20,963
Outlays	7,251	7,403	7,591	9,092	11,036	12,026	14,137	16,413	18,148	19,379	20,122

Note: Figures through fiscal year 1989 are actuals, fiscal year 1990 is an estimate, fiscal year 1991 is the Budget Request, and fiscal year 1992 onward are FY 1991 President's budget projections which may not fully fund the program described in the text.

(all in 1988 dollars) − or more than 4 percent average real annual growth. (See Table 5.1.) Table 5.1 also shows the costs of several other more ambitious program plans.

The Bush Administration Additions

The Bush administration appears to have become an even stronger booster of space spending than the Reagan Administration was. So far, the Bush administration has added dramatic new programs to the NASA "core program", including the Moon/Mars Mission − estimated to cost more than $540 billion in real fiscal year 1991 dollars over a 35-year period − and the Earth Observation System − estimated to cost more than $17 billion in real fiscal year 1991 dollars by the year 2000. The fiscal year 1991 Bush program's 5-year budget request is displayed in Table 5.2. This will require almost 12 percent average annual growth − or approximately eight percent real growth. A still larger estimate comes from the numbers in the CBO study which indicate greater than 10 percent average real annual growth. Thus we have set the scene: The core program is planned to grow, and the President (and others) support new programs which will add even more growth to the budget.

The Problem

Why is this a problem to budget makers in the Congress? After all, NASA accounts for approximately one percent of the Federal budget and less than 0.25 percent of GNP, and the kinds of numbers involved with implementing the "core program" and the Bush proposals would take these numbers to somewhere between 2 to 3 percent of the Federal budget and less than 0.5 percent of GNP. Compare this to fiscal year 1966, NASA's peak year in real spending levels: The NASA budget that year was approximately $23 billion in fiscal year 1990 dollars, which represented approximately four percent of the Federal budget and 0.8 percent of GNP. So why shouldn't we be able to fund NASA at those levels again?

The answer to this question requires a description of the budget process, how it works in detail, and why it is difficult in the budget climate we now face to make this kind of *redirection* of resources from other programs and into NASA.

As a starting point consider a thumbnail sketch of the Federal budget. It is convenient to think of the Federal budget in the following four expenditure categories: defense, entitlements, interest, and nondefense discretionary programs. The categories are not perfect — some programs could fall into two or three categories — but they are good enough to describe the general shift in Federal spending since 1980. And they are categories used in budget negotiations. A description of the categories follows:

Defense — This includes the expenditures of the Defense Department, and the atomic weapons expenditures of the Department of Energy.

Entitlements — These are expenditures made pursuant to permanent law; that is, they do not require annual appropriation or annual legislative approval for their existence. For example, this category includes Social Security, Medicare, Federal retirement programs, and agricultural price support expenditures, among others.

Interest — This category represents the interest that must be paid on prior-year Federal government borrowing. Because the total Federal debt has more than tripled since 1980, this has been the fastest growing expenditure category. (Federal debt that totaled $914 billion at the end of fiscal year 1980, totaled $2.866 trillion by the end of fiscal year 1989 and was expected to reach nearly $3.539 trillion by the end of fiscal year 1991.)

Nondefense discretionary programs — This comprises everything else: education, energy, Labor Department, NASA, NSF, NIH, Commerce Department, regulatory commissions and other expenditure programs for which funds are *appropriated on a year-to-year basis.*

Table 5.3 shows what has happened to these expenditure categories over time:

Defense — has gone up from $134 billion in fiscal year 1980 and $157 billion in fiscal year 1981, to $304 billion in fiscal year 1989. This was the so-called Reagan defense buildup.

Entitlements — that were $320 billion in fiscal year 1981 were $544 billion in fiscal year 1989.

Interest — that was $68.7 billion in fiscal year 1981 was $169 billion in fiscal year 1989.

In the meantime, nondefense discretionary programs peaked at $171 billion in fiscal year 1981, actually went *down* to $156 billion in fiscal year 1983, and then increased to $191 billion in fiscal year 1989.

A few points can be made about these data:

- Nondefense discretionary program expenditures have grown by less than 12 percent in nominal terms from their 1981 peak.
- In real terms — that is, after adjustment for inflation — they have been cut by about 15 percent. And as a percent of the total Federal budget, they have declined from 25 percent in 1980, to less than 16 percent in 1989.
- Of course, this decline is true of the category as a whole, but within this category, certain programs have fared well — such as NIH, NSF, and even NASA — while others have been cut to the bone: for example, the Synthetic Fuels Corporation and General Revenue Sharing have been eliminated.

Table 5.4 is another way of looking at the same data. It presents the same story, but as a percentage of GNP rather than in nominal dollar terms. You can see that nondefense discretionary programs peaked at 5.9 percent of GNP in 1980 and then proceeded to decrease to 4.4 percent in 1985, and in 1989 are estimated to represent 3.7 percent of GNP.

In sum, the funding available for discretionary programs is shrinking, while some individual programs are growing.

The Budget Process

Why is this story so important? Because NASA spending falls in this nondefense discretionary program category. And NASA spending — like all discretionary spending — is appropriated on a year-to-year basis. It is also the category of spending that most appropriations subcommittees are involved with. As will be seen below, eleven of the thirteen subcommittees of the appropriations committees of each House compete among themselves for this pot of funds.

How does Congress decide what to spend, overall, on a year-to-year basis? This is determined by the yearly budget process and, ever since

Table 5.3. Outlays for Major Spending Categories,
Fiscal Years 1962-1989 (In billions of dollars)

Fiscal Year	National Defense	Mandatory Spending[1]	Nondefense Spending[2]	Net Interest	Offsetting Receipts	Total Outlays
1962	52.3	30.7	23.9	6.9	-7.0	106.8
1963	53.4	33.2	25.1	7.7	-8.1	111.3
1964	54.8	34.4	29.0	8.2	-7.8	118.5
1965	50.6	34.7	32.3	8.6	-8.0	118.2
1966	58.1	37.5	38.1	9.4	-8.5	134.5
1967	71.4	45.3	40.8	10.3	-10.3	157.5
1968	81.9	52.3	43.6	11.1	-10.8	178.1
1969	82.5	58.5	41.1	12.7	-11.1	183.6
1970	81.7	66.2	45.0	14.4	-11.6	195.6
1971	78.9	80.6	50.1	14.8	-14.2	210.2
1972	79.2	94.2	56.1	15.5	-14.2	230.7
1973	76.7	110.2	59.6	17.3	-18.1	245.7
1974	79.3	124.4	65.4	21.4	-21.3	269.4
1975	86.5	156.4	84.7	23.2	-18.5	332.3
1976	89.6	182.8	92.4	26.7	-19.7	371.8
1977	97.2	196.5	107.2	29.9	-21.6	409.2
1978	104.5	216.3	125.5	35.4	-23.0	458.7
1979	116.3	234.2	136.3	42.6	-26.1	503.5
1980	134.0	277.2	157.6	52.5	-30.3	590.9
1981	157.5	320.4	170.8	68.7	-39.2	678.2
1982	185.3	356.0	156.6	85.0	-37.2	745.7
1983	209.9	398.8	156.0	89.8	-46.1	808.3
1984	227.4	394.7	163.9	111.1	-45.3	851.8
1985	252.7	437.3	174.9	129.4	-48.0	946.3
1986	273.4	454.8	173.2	136.0	-47.0	990.3
1987	282.0	472.4	165.1	138.6	-54.2	1003.8
1988	290.4	502.7	177.2	151.7	-58.0	1064.0
1989	303.5	543.6	191.0	168.9	-64.2	1142.9

Source: Congressional Budget Office

[1] Entitlements and other mandatory spending

[2] Discretionary Spending

1986, its interaction with the Gramm-Rudman-Hollings (GRH) process. GRH was enacted first in November 1985 (GRH I), amended in November of 1987 (GRH II), and amended most recently in November, 1990. For simplicity the following description covers only the House of Representatives, but a similar process occurs in the Senate.

In January of every year, the President sends up his budget which is reviewed by the Congress, and in early March the Budget Committee starts preparing its version of the Budget Resolution forthe fiscal year that begins on October 1. The Budget Resolution has to come up with a way of cutting the deficit, i.e., lowering it from the so-called "CBO current policy baseline" in a way that will lead to a budget which, if enacted, will meet the desired deficit target for that year. Under the GRH I and II processes, the deficit targets were placed in law, and thus were exogenously determined. If that target deficit was not reached as of the beginning of October, "sequestration" would be triggered, causing automatic across-the-board cuts in most appropriated accounts. That is, the projected deficit had to be judged to meet the GRH target or else automatic, proportional cuts would be made, sufficient to achieve the mandated overall deficit target. In 1990 GRH was further amended to replace the single annual deficit targets with a system that placed annual ceilings on three categories of discretionary spending, and a pay-as-you-go system for mandatory spending programs.[5]

The Budget Committee makes assumptions about the cuts that can be made in all appropriated spending accounts, the cuts that can be made in entitlement programs, and the increases that can be made in taxes and user fees to reach the target deficit. The cuts that requirechanges in permanent law (for example, entitlement law changes and tax and user fee increases) are generally "reconciled"[6] (the committees of jurisdiction are directed in the Budget Resolution to come up with the savings) and the remaining spending amounts are assigned to other spending committees.

The biggest such spending committee is the Appropriations Committee. The amount of budget authority and outlays assigned to each Committee of the House by the Budget Resolution is referred to as the "602(a) allocation" after the section of the Budget Act that requires it.

The next step in the process is when each committee reports back to the full House of Representatives how it has allocated *among its subcommittees* the money it has been given by the Budget Resolution.

This second allocation of funds, among subcommittees of each committee, is referred to as the "602(b) subdivision", after the section of the Budget Act that requires that it be prepared. There are thirteen subcommittees of the Appropriations Committee, and the competition among the subcommittees is heated. The subdivision has to be ratified by the committee as a whole, but in practice the decision is made by the thirteen subcommittee chairmen led by the chairman of the full committee, who usually chairs one of those subcommittees.

Each Appropriations subcommittee has to stay within its 602(b) subdivision, or it will encounter virtually automatic and probably fatal procedural problems when it brings its appropriations bill to the floor of the House.[7] Therefore, those interested in NASA spending automatically should be interested in the 602(b) subdivision for the VA-HUD (Veterans Affairs, Housing and Urban Development and Independent Agencies) Appropriations Subcommittee, which is responsible for NASA funding, among other programs. To put it differently, the VA-HUD subcommittee cannot fund projects with money it does not have.

Similarly, the full appropriations committee of each House is limited in what it can give its subcommittees by how much it receives in its 602(a) allocation from the Budget Resolution. To repeat, those interested in the VA-HUD 602 (b) subdivision automatically should be interested in how much money is being assumed for nondefense discretionary programs in the Budget Resolution.

It is clear that the total amount made available for nondefense discretionary spending is extremely important for determining how much the VA-HUD subcommittee will get and, in turn, how much that subcommittee can use to fund NASA programs. Table 5.5 displays the 602(b) subdivision made by the House Appropriations Committee using the House-passed Budget Resolution for Fiscal Year 1991 and shows how it compares to the Senate Appropriations Committee's 602(b) subdivision.

Note that the political dynamics change at every stage of the process. Supporters of NASA programs, at the Budget Resolution formulation level, generally should be expected to support increases in nondefense discretionary programs, which means they should support — all other things being equal — decreases in entitlement programs, increases in user fees and taxes, and decreases in defense. However, it is important to realize that many advocates of space spending are defense contracting companies which many times have greater interest in defense spending than in space spending.

Once the 602(a) allocation has been set, supporters of various nondefense programs who had been allied *through* the Budget Resolution stage, will now compete against each other when setting the 602(b) subdivisions *among* subcommittees of the Appropriations Committee.

Finally, once the 602(b) subdivisions are set, supporters of programs in a given subcommittee — who had been allies up to this stage — will now compete against each other within that subcommittee. So advocates of space spending would now be competing against advocates of Veterans, HUD, EPA and NSF funding.

How has this process treated NASA? It has been better than one might have expected, given the budget pressures we have been discussing. After all, NASA funding has increased from $7.8 billion in fiscal year 1986 to $12.3 billion in fiscal year 1990 — nearly a 58 percent increase in four years, or roughly 12 percent average annual growth — pretty good in an era of Gramm-Rudman limits. On the other hand, this is far below the amounts requested by NASA to support the program it believes it has been charged with implementing.

How was it possible for Congress to give NASA these kinds of increases while discretionary spending budgets increased by far smaller proportional amounts? After all, the bill that funds NASA also funds the Departments of Veterans Affairs and Housing and Urban Development, the Environmental Protection Agency, and the National Science Foundation, among other agencies and programs, and many of these also have strong demands for funding growth. (For example, the President for several budget years has promised to double NSF's budget over five years.) One of the reasons is that the supporters of NASA have been creative in finding ways to increase NASA funding which would cause as little competition to other funding requirements in the VA-HUD bills as possible.

For example, in fiscal year 1987, Congress appropriated $2.1 billion in budget authority to replace the Challenger shuttle, but it did so in an appropriation bill that took the money from funds that previously had been allocated for defense purposes. In 1986, after the Challenger disaster occurred, it became apparent that another shuttle vehicle would be required if the shuttle system was to be continued. However, the Reagan administration did not make a supplemental appropriation request (of $250 million) until August. Also, the VA-HUD Appropriations Subcommittees had not included funding for this purpose in their respective bills. However, NASA supporters in the

Table 5.4. Outlays for Major Spending Categories,
Fiscal Years 1962-1989 (As a percentage of GNP)

Fiscal Year	National Defense	Mandatory Spending[3]	Nondefense Spending[4]	Net Interest	Offsetting Receipts	Total Outlays
1962	9.4	5.5	4.3	1.2	-1.3	19.2
1963	9.1	5.7	4.3	1.3	-1.4	18.9
1964	8.7	5.5	4.6	1.3	-1.2	18.8
1965	7.5	5.2	4.8	1.3	-1.2	17.6
1966	7.9	5.1	5.2	1.3	-1.2	18.2
1967	9.0	5.7	5.1	1.3	-1.3	19.8
1968	9.6	6.2	5.1	1.3	-1.3	21.0
1969	8.9	6.3	4.4	1.4	-1.2	19.8
1970	8.2	6.7	4.5	1.5	-1.2	19.8
1971	7.5	7.6	4.7	1.4	-1.3	19.9
1972	6.9	8.2	4.9	1.3	-1.2	20.0
1973	6.0	8.6	4.7	1.4	-1.4	19.2
1974	5.6	8.8	4.6	1.5	-1.5	19.0
1975	5.7	10.3	5.6	1.5	-1.2	21.8
1976	5.3	10.8	5.4	1.6	-1.2	21.9
1977	5.0	10.2	5.5	1.5	-1.1	21.2
1978	4.8	10.0	5.8	1.6	-1.1	21.1
1979	4.8	9.6	5.6	1.7	-1.1	20.6
1980	5.0	10.4	5.9	2.0	-1.1	22.1
1981	5.3	10.7	5.7	2.3	-1.3	22.7
1982	5.9	11.3	5.0	2.7	-1.2	23.8
1983	6.3	12.0	4.7	2.7	-1.4	24.3
1984	6.2	10.7	4.4	3.0	-1.2	23.1
1985	6.4	11.1	4.4	3.3	-1.2	23.9
1986	6.5	10.9	4.1	3.3	-1.1	23.7
1987	6.4	10.7	3.7	3.1	-1.2	22.7
1988	6.1	10.5	3.7	3.2	-1.2	22.2
1989	5.9	10.6	3.7	3.3	-1.2	22.2

Source: Congressional Budget Office

[3] Entitlements and other mandatory spending

[4] Discretionary Spending

Senate noticed that the Defense Appropriations Subcommittee was in the process of reporting its bill which had used *all* of the outlay allocation it had been given, but *not* all of its budget authority allocation. These NASA supporters prevailed on the Defense Appropriations Subcommittee to use some of that "excess" budget authority to fund the replacement shuttle vehicle in the Defense Appropriations bill. Perhaps as important, the Congress enacted in 1986 an omnibus fiscal year 1987 appropriations bill containing *all* thirteen subcommittee bills. This made it possible, when in conference, to shift this money from the Defense Appropriations bill to the VA-HUD bill without encountering a number of procedural problems relating to the 602(b) allocation for each subcommittee. Note also that no distinction existed at that time between defense and domestic funds similar to the separate limits on each category of spending that are in effect for fiscal years 1991-93. Such limits would have made this transfer impossible.

In fiscal year 1989, the Congress made a substantial part of the appropriation for the Space Station ($515 million) available for obligation, only after May 15, 1989, subject to release by whomever was the next President. The effect was to substantially reduce in fiscal year 1989 the *outlay* costs of the Space Station program, leaving more outlay room within the subcommittees' allocation to fund other necessary programs. In fiscal year 1990, the Congress made substantial reductions in NASA outlays by delaying the obligation of several accounts, but at the cost of pushing the same outlays into future years. Other creative techniques, as well as transfers from DOD were implemented, but this kind of creative funding maneuver will become increasingly difficult over time. The extent to which the VA-HUD subcommittees' 602(b) subdivisions will continue to grow relative to the other 12 appropriations bills is also unclear. The upshot is that it may become relatively more difficult to fund rapidly growing NASA budgets out of overall funding pots that are growing at only the general inflation rate, if that fast.

Future Prospects

Could the prospects for NASA funding improve? Yes, of course that is possible; however, it may not be likely. Through 1995, GRH sets limits on discretionary spending that grow increasingly tight. NASA

Table 5.5
FY 1991 Budget Resolution Assumptions (Discretionary Funds)

Appropriations Committees	Senate 302(b) Subdivision		House 302(b) Subdivision		Difference	
	BA	O	BA	O	BA	O
Commerce/Justice/ State/Judiciary	19335	18289	19345	18753	10	464
Defense	263324	276078	262804	276054	-520	-24
District of Columbia	570	580	570	580	0	0
Energy/Water Development	20900	19900	20900	19900	0	0
Foreign Operations	14678	12626	15100	13200	422	574
Interior	12800	11925	12800	12000	0	75
Labor/Health & Human Services	50408	53208	52753	54008	2345	800
Legislative Branch	2268	2155	2268	2250	0	95
Military Construction	7980	8666	8500	8700	520	34
Rural Development	10167	9476	10167	9559	0	83
Transportation	12700	2900	12600	29200	-100	200
Treasury/ Postal Service	11550	10300	11550	10300	0	0
VA/HUD/IND	61320	58545	63400	59175	2080	630
GRAND TOTAL DISCRETIONARY	488000	510748	492757	513679	4757	2931

funding prospects, therefore, depend on the depth of commitment of the administration and of NASA's supporters in the Congress, their skill in coalition building, and their willingness and ability to trade for support in the political process. They also strongly depend on the strength of public support for these programs, and the public's support for space relative to its interest in other priorities.

How NASA fares will also depend on what happens to Gramm-Rudman-Hollings and the reductions Congress is able to make in other programs on a longer-term basis (such as defense, and medical and agricultural entitlements, among others) and the ability of Congress to raise user fees and taxes. Future changes to the GRH process will be particularly important: for instance, whether a new GRH mandates more deficit reduction, or whether it continues to establish separate ceilings on the different categories of discretionary spending, precluding transfers between defense and nondefense programs. In any event, it will be very difficult for NASA over a prolonged period to sustain real annual growth increases of 10 percent in a nondefense discretionary budget which may experience somewhere between plus or minus three percent real growth. Higher growth rates for NASA's budget will depend on a continuing ability of Congress to make reductions in entitlement and defense spending.

Is There Anything Else That Could Be Done?

Some in Congress have advocated the adoption of "pay as you go" budgeting techniques. These require that new spending initiatives be paid for with new taxes, or user fees. Some of the space initiatives presently being discussed are of an infrastructure nature and may have sufficient public appeal to generate the political support needed to enact new taxes or user fees to pay for themselves. These receipts could be deposited in a Treasury account and directly used as a source of funding for some of these programs, somewhat similar to the way trust funds have been used to finance other programs in the past. By essentially establishing a self-financing program, it would be possible to undertake these initiatives without the fear that they would crowd out other domestic spending priorities, and without the heat of that competition. This approach may become of increasing interest, unless the overall Federal budget picture changes in some unforeseen, dramatic way.

Notes

1. Michael L. Telson serves as a Senior Budget Analyst, Committee on the Budget, U.S. House of Representatives. This paper represents his own views and not necessarily those of his employer.
2. NASA's "core program" was defined by CBO as the program that served as the basis for formulating NASA budgets since the mid 1980's, when the Reagan administration proposed its general outlines. Broadly defined, it includes NASA's manned space station, an aggressive space shuttle program and a number of space science missions then envisioned, but did *not* include the Earth Observation System (Mission to Planet Earth) or a Moon-Mars exploration effort which were adopted by the Bush administration.
3. "Nondefense discretionary" refers to spending which is appropriated by Congress on a yearly basis (hence discretionary) for programs with other-than-defense purposes.
4. Congressional Budget Office, *The NASA Program in the 1990's and Beyond*, USGPO, Washington, D.C., May 1988.
5. In November 1990, the President signed into law the Budget Enforcement Act of 1990 (title XIII of the Omnibus Reconciliation Act of 1990). This act provides a new set of budget procedures which will be in effect during fiscal years 1991-95. The most significant provisions of the Act (for our purposes here) specify *separate*, aggregate totals for the following three categories of discretionary spending: defense, international and domestic. These separate spending pots and ceilings are in effect during fiscal years 1991-93. In fiscal years 1994 and 1995, only the *aggregate* ceilings on *all* discretionary spending remain in effect. These ceilings are enforced by *separate* automatic sequesters that go into effect whenever the spending in either category is estimated to exceed its budget authority or outlay ceiling. This means that for fiscal years 1991-93, *unless* a budget resolution *further limits* spending, or new law is enacted changing the budget procedures, the section 602(a) allocations are in large part determined. It also means the same thing for fiscal years 1994 and 1995, but with larger discretion for the appropriations process to shift money among these spending categories.

The Budget Enforcement Act of 1990 (BEA) also creates a new "pay as you go" procedure to finance new permanent spending programs financed with the proceeds of new taxes, user fees, or from savings in *existing* mandatory spending statutes. The BEA also allows for a return to fixed-deficit targets beginning in 1994, if the President so chooses.

The important point to note is that the BEA, in combination with Budget Act procedures, does not change the implications of this paper. The remainder of this paper applies to the new budget procedures as well as to the previously existing or future processes, as now anticipated. The ultimate budget-making reality in a Gramm-Rudman environment is the zero-sum nature of the process.

6. For example, the House-passed Congressional Budget Resolution in 1990 directed the House Energy and Commerce Committee and the House Interior and Insular affairs Committee, as well as the Senate Environment and Public Works Committee, to report a bill that would contain almost $300 million in yearly savings from the enactment of full cost recovery user fees for the Nuclear Regulatory Commission (NRC). This involved giving the NRC the legal authority to charge its licensees fees

sufficient to cover the full costs of operating the NRC. This provision ultimately was enacted into law in the Omnibus Budget Reconciliation Act of 1990.

7. That is, any Member can object to an appropriations bill that is over its allocation by raising a "point of order" which, under the Rules of the House, will be sustained without a vote, thus killing the bill.

Chapter 6
DECISION MAKING AND ACCOUNTABILITY IN THE U.S. SPACE PROGRAM: A PERSPECTIVE

Jack D. Fellows

> THERE WAS THINGS WHICH HE STRETCHED, BUT MAINLY HE TOLD THE TRUTH.
> —MARK TWAIN

Introduction

Twain is a beloved American author, partly because of his wit and his gift of capturing the very spirit of his characters, and in fact society, in a few simple words. An example of this gift is the epigram above in which of one of his characters, Huck Finn, describes Mark Twain. In one sentence, the reader learns that Huck felt Twain was basically a person of pure motive, yet at the same time one should maintain a certain skepticism concerning the details of his claims.

Similarly, there is something in this characterization reflected in the history of our nation's space program. Through seemingly forthright technological and visionary leadership, our nation has accomplished some spectacular space events like manned Moon missions, the Space Shuttle, and the unmanned Voyager "Grand Tour" of the Solar System. Yet, at the same time it is curious why there seems to be so little continuity between NASA programs (e.g., the Apollo missions and the Human Exploration Initiative, Skylab and Space Station, the Saturn V rocket and new heavy lift launch vehicles). Given this lack of continuity, the Challenger accident, the recent public criticism of NASA management, and the many problems facing NASA today with the Shuttle, Space Station, and Hubble Space Telescope, one wonders whether the public should have adopted a more "Huck-like" skepticism toward our space program planning process.

The overall purpose of this paper is to examine how the U.S. civilian space program has been shaped by the incentives and accountability mechanisms that influence decision making. The specific tasks are to address the following questions:

- How are program decisions driven by *de facto* incentives, and what real incentives do decisionmakers see?
- How does program accountability manifest itself, which aspects of programs are appraised and which are not, by whom, and with what effect?

These are very broad questions and involve hundreds of thousands of individuals and programs.

Scope of the Paper

Because of this breadth, the scope of the discussion will be limited to several specific levels of decisionmakers and programs. The incentives and level of accountability differ radically among individual decisionmakers and programs. The paper will focus on decisionmakers in industry, NASA, the Executive Office of the President, and the Congress; for three specific categories of NASA space programs: manned missions (e.g., Shuttle, Station, human exploration), large unmanned programs (e.g., billion dollar astronomy facilities) and small unmanned projects (e.g., Explorers, suborbital flights, research and analysis, etc.). Within this scope the paper will attempt to identify possible changes that might contribute to a more effective and sustainable decision making process.

U.S. Civilian Space Program

Regardless of one's opinion of NASA, the technical success the agency has enjoyed over the past three decades is impressive. This success is largely due to the technical competence and leadership embodied in the NASA workforce. This spirit and professionalism became a formal policy in Administrator James Beggs' January 1982 *Principles of Project Management*. The following is an excerpt from that document and highlights NASA official policy on program decision making and accountability:

> While high risk is inherent in the work done by NASA, every effort will be made to understand and quantify this risk before seeking Office of Management and Budget and Congressional commitment to a project. The desire to obtain project approval or to proceed with project implementation should not be permitted to interfere with the adequate and comprehensive studies and cost analyses necessary to define the risk so as to minimize the potential for cost overruns and schedule slips.

A central point in this statement is that project approval should not be sought until the risk of cost overrun and schedule slippage has been minimized. Within NASA and industry there are significant individual incentives (e.g., professional advancement, personal pride, award fees, etc.) and accountability mechanisms (e.g., program milestones, external reviews, audits, etc.) imposed to reach the goal outlined in Administrator Beggs' policy. When cost-overruns and schedule slips do occur, they may indeed stem from unforeseeable technological problems associated these high-risk space projects. However, it is just as likely that these problems were either created or amplified by *de facto* incentives and the lack of enforceable accountability.

For example, from time to time one hears complaints from industry or the science community about what a financial or scientific mistake it was to pursue a particular approach to the implementation of a NASA program. Yet when asked why they weren't more vocal during the planning phase, the response is often that there is little incentive to challenge NASA when that program represents the only flight opportunity for decades. What other incentives have driven these conspirators to remain silent and why haven't accountability mechanisms been effective to reverse their impact?

In reality, the influence of these *de facto* incentives have been even further enhanced by the extremely competitive funding environment. Thus, those with oversight responsibilities must remain very suspicious of decisionmakers' incentives and accountability.

Incentives and Accountability

In addition to the incentives of individuals mentioned earlier, the are many other broader incentives that justify investments in a national space program, including:

- a President's desire to involve the nation in an exciting and technologically challenging space goal (e.g., Kennedy-Moon,

Nixon-Shuttle, Reagan-Station, and Bush-Human Exploration), and
- a strong belief within the Administration and the Congress that investments in the space program are important and provide an appropriate return on the investment for the nation, including:
  – satisfaction of operational needs (e.g., weather data),
  – technological advancement,
  – the creation of new knowledge,
  – the training of scientists and engineers, and
  – the creation and realization of commercial opportunities in space.

The accountability of these investments is also critical. Ideally, incentives and associated accountability mechanisms should result in the establishment of a flexible and evolutionary long-term space program with many clear, near-term benefits. The pace and needed resources of the program should also be scaled and integrated to fit within the expected resources and competing needs of the nation's economy.

Unfortunately, several *de facto* incentives have evolved over the last several decades that have had a decisively negative impact on the prospects for effective, long-term space program planning. These de facto incentives include:

Realistic planning vs. vision. Many program decisions are driven by the pursuit of grand visions (e.g., man on the moon, Shuttle, Space station, human exploration, multi-disciplinary science missions, etc.) that have been enthusiastically offered to and enthusiastically accepted by decisionmakers. The resource needs and technological requirements of these vision are often grossly understated, in conflict with other national needs, and difficult to sustain over the long-term. However, because of the tremendous political appeal of the grander visions, there are few incentives to examine less visionary but more affordable approaches. The decisionmakers "buy into" the vision only to leave their successors with a significant future funding and credibility problem. The relatively short-term tenure of Presidents and the annual appropriation process supports the proliferation of this practice.

Cost estimates. As long as cost is a predominant criteria for mission or contractor selection but not controlled during program execution, there will be a strong incentive for cost estimates to be significantly underbid. While cost should always be an important factor, technical approach, schedule, program output, appropriate reserves, performance, quality, technical competence, personnel, and the management plan are frequently more important long-term parameters in selections.

Mission requirements. Because it is difficult to define budget constraints in the preliminary definition stage of a project, there is a strong incentive to approach the development of mission requirements (e.g., scientific, technical, safety, etc.) with little regard for budget. The result is normally a set of requirements that both are inconsistent with available resources and become sacrosanct with the mission's designers.

Risk avoidance. There is a strong incentive to over-design projects to minimize the chance of failure, despite the fact that in many cases this is a very costly and counterproductive practice. A former NASA Shuttle engineer once complained privately that he estimated it would cost roughly $100,000 to space-qualify a common building brick. There is little incentive within the NASA community to take high-risk, innovative approaches to space missions.

Infrastructure. Because of the constrained budget environment, there is an increasing pressure to maintain NASA research centers and on individual decision-makers to divert space program funds for their own constituents, rather than to pursue potentially more competitive, innovative, and higher-return investments.

Astronauts. Manned missions (e.g., Apollo, Shuttle, etc.) have become synonymous with maintaining adequate political support for NASA programs. It is unclear whether the benefits outweigh the significant additional costs associated with human involvement.

Constituencies. The threat of competition for resources has forced very odd marriages (e.g., a polar orbiting platform as part of the Space Station, sensitive microgravity materials labs on a dynamic Space Station, multi-disciplinary science projects, etc.). The incentive to accommodate all possible constituents may increase political support but normally reduces the overall performance; i.e. achievement of the individual requirements is compromised. In addition, there has been an incentive to support as many preliminary mission definition studies as possible to satisfy the many interested and demanding space constituencies (e.g., science, technology, human exploration, etc.). This spreads false expectations and distributes very scarce funding resources to projects that are unlikely to be undertaken in the near future.

Large vs. small programs. Small projects can take roughly the same amount of energy to initiate as large and complex projects, without the potential for an equivalent payoff or positive career exposure. Thus there may be little incentive to support small programs. For similar reasons, there are more incentives to initiate new programs instead of adequately supporting existing programs.

The Impact of De Facto Incentives and Ineffective Accountability

Over the past three decades NASA has proposed the nation's civilian space agenda and with only minor adjustments (excepting the clamp-down after Apollo) this agenda has been blessed by the President and approved by the Congress. The process begins with a NASA proposal undergoing a variety of reviews (i.e., NASA program reviews, external scientific advisory committee reviews, and external technical non-advocacy reviews). Program decisions are made at every level and program accountability mechanisms are introduced throughout the process (i.e., milestones, program evaluations, performance appraisals, etc.). Formal approval is then sought from the President and the Congress.

Throughout this process there are few incentives to promote affordable, low-tech, and evolutionary space missions. There is also very little effective accountability. A few examples will quickly illustrate these two points.

Manned missions. The success of the Apollo mission provided NASA with the technological capability and public credibility to undertake even more challenging manned projects. Unfortunately, there have been few incentives to develop an affordable and evolutionary human exploration program. Despite the fact that cheaper and well-integrated missions ought to be easier to support, we continue to build very expensive and loosely connected programs. Along the way, we have discarded important technologies (e.g., Saturn V heavy lift launch vehicle, Skylab), produced a Shuttle that is far from an economical and reliable system, and spent billions of dollars defining a Space Station that appears to have less and less to offer. Why has this happened? There is no simple explanation to this question, but let us look at some of the incentives that influenced these outcomes.

Expensive spacecraft with extensive technological backup systems can significantly reduce the risk of failure. A NASA Program Manager who successfully executes an expensive, multi-redundant spacecraft is likely to be significantly rewarded. On the other hand, the failure of a innovative, cheap, high-risk spacecraft is likely to destroy a career. Given, these two choices, which approach would you chose? This incentive to build complex and expensive programs is further amplified by many of the other *de facto* incentives:

— As the cost of spacecraft increase, fewer can be built. Thus, the incentive to build even larger and multi-purpose spacecraft intensifies in an attempt to satisfy over-estimated mission requirements, support NASA infrastructure, and broaden a project's political base.
— There is a strong incentive to undertake exciting state-of-the-art, high-profile manned projects despite the fact that they may have little or no realistic long-term strategy (i.e., man-on-the-moon, Shuttle, Station, etc.). Scientists and engineers are strong advocates for this high-tech approach since there may be little personal incentive to invest their careers in the development of low-tech operational infrastructures.
— There continues to be significant appeal to including astronauts as a central part of missions despite the fact that the health and safety requirements of manned spaceflight send a program's costs soaring.

The result of this process is a prohibitively expensive set of disjoint programs that support many requirements but satisfy few. They tend to be disjoint because their individual price tags are so high that it is unrealistic to propose them as an integrated set of long-term objectives (i.e., an affordable space infrastructure that supports a reasonable long-term human exploration effort).

For example, the original performance claims of the Shuttle made it appear that there was little that the Shuttle could not do (i.e., science, technology demonstrations, cargo vehicle, personnel carrier, etc.) and that it would do it inexpensively (i.e., 60 flights per year, $100 per pound to low earth orbit). Despite the disastrous outcome of this policy, we even became convinced that Shuttle should be the U.S.'s primary launch vehicle. What resulted, largely because of the *de facto* incentives, was a shuttle that is so complex and expensive that it strains the budget and NASA operational capability even when launched less than a dozen times per year. Many significant and painful lessons should have been learned from the Shuttle program. The most important being that a less complex, cheaper, and more operational system should have been sought in the design of the Shuttle. It is very difficult to see how the current Space Station reflects this important principle. The absence of these principles in the Space Station program underscores the power of the *de facto* incentives and the lack of effective accountability.

The Space Station was originally designed to support nearly every conceivable space activity, including permanently manned capability, materials research, life sciences, earth observations, physics and astronomy, planetary exploration, commercial activities, robotics, and spacecraft servicing. All this and more, within a decade, and for only $8 billion. It was the perfect candidate for Presidential support, exciting, visionary, high-tech, and could be extended to the international community as a significant foreign policy tool. It was spread throughout NASA, the science community, and the aerospace community to broaden the involved constituencies and to support NASA infrastructure. Contractors made very low bids to remain competitive but were privately quite concerned that the real costs were significantly greater ($15-20 billion). The current Space Station is not affordable, nor does it "fit" into any long-term plan. For what is it really needed relative to its price tag? If there is to be a long-term human exploration program in the future, the most attractive feature of a space station would be a platform for long-duration life sciences research. Anything else conceivably can be done better and probably cheaper via a manned-tended approach.

There is limited incentive for the Executive Office and the Congressional staff to oppose or even to question rigorously these large manned projects. They are extremely complex and on many occasions have been "pre-approved" through the strong endorsement by the President or the Congress while the project is still in the pre-development phase. Rarely are there any technically competent alternatives to chose between.

It is reasonable to ask why there hasn't been anyone held accountable for all of these costly and disjoint decisions? Yes, many people were accountable and if they were involved again the result would likely be the same. The *de facto* incentives provide strong disincentives to radical change. The Shuttle and the Station were supported by both the President and the Congress as grand visions of space achievement. In truth, they are more short-term technology demonstrations and foreign policy tools than rational components of an affordable, long-term space program.

The visionaries are normally long gone before the program becomes a reality. Thus, their successors are saddled with actually paying for and accomplishing the mission. High-level endorsements make effective accountability nearly impossible. Outside of NASA, accountability is limited to what Executive Office and Congressional staffs can challenge (i.e., schedule, content, resource needs, management approach, etc.) given the tremendous management and technological complexity of these programs. The primary focus of these non-NASA reviews is to get the most acceptable program given the fiscal constraints and political realities. Significant departures from the original NASA implementation approach are normally minimal, unless the top-level political support begins to waver. For example, throughout the 1980s, changes were minimal on the Space Station program despite the evidence that the cost and capabilities were grossly under and overestimated respectively. Finally, the constrained budget environment and the lack of realistic and clearly defined long-term requirements precipitated a much needed and critical review of the Space Station.

The bottom line is that for these high-profile manned missions there is almost no accountability. For the most part, cost overruns due to management problems, schedule delays, or accidents were paid as the price of high-risk, politically motivated programs. The only visible effort at accountability is the replacement of a NASA program manager. On the Space Station, rapid turn-over of program managers further eroded accountability. Each year, problems were attributed to the previous year's management team. Of course, there were a multitude of Inspector General, special commission, General Accounting Office, Office of Technology Assessment, and other reports, but these are usually too late to be an effective input into the decision-making process. These external reviews are usually initiated to better understand why a problem occurred, not as a planning exercise. It was unusual when a review was focused on questioning the fundamental merit of a NASA proposal.

Large unmanned programs. Given the existence of the various non-NASA planning efforts for unmanned science programs (e.g., advisory groups, merit review, Academy Reports, etc.), there are usually few questions from the Executive Office or Congress about the scientific merit of these proposals. Hubble Space Telescope (HST), the Advanced X-ray Astrophysics Facility (AXAF), the Comet

Rendezvous-Asteroid Flyby/Cassini missions, and the Earth Observing System (EOS) have all undergone these reviews and the incentive to do good and important science is compelling. However, in most cases these reviews focus primarily on the science requirements and not on the overall technological and management implementation package, *nor on the budget.*

Unlike the manned programs, the President and the Congress have provided much less critical input and generally have supported the unmanned programs. Program decisions have been made predominately by NASA program managers and senior officials. Nevertheless, the program decisions are just as influenced by the same *de facto* incentives. Most of these projects have become very costly, multi-disciplinary, and complex programs that are spread throughout the same communities as the manned programs. And for the reasons cited above, the science, contracting, and other communities have remained relatively silent about this situation.

The public justification has been that the natural extension of the science requires this complexity. Yet, the complexity and cost increases in themselves limit the science through fewer research and training opportunities. Hubble, AXAF, and other large unmanned programs have the same potential for disaster as our sole reliance of the Space Shuttle for access to space. In some instances, it is questionable if a proper balance between investment, risk, and return has been maintained.

At this program level, accountability tends to be limited to the NASA program manager. Executive Office and Congressional staff have been extremely respectful of the merit review process, the need for NASA to have a balanced program (i.e., research vs. operations), and NASA's implementation approach. Careful scrutiny of the year's funding requirements and progress is undertaken, but the outcome is usually limited to juggling the funding schedule. Much of this accountability is maintained within NASA by senior NASA officials forcing program managers to live within their annual budget. However, when problems occur such annual discipline merely forces the resolution of the problems into later years, thus increasing total program costs.

AXAF and EOS may be the first exceptions to this historical process. Most of even the largest unmanned projects are below the typical Shuttle/Station budget threshold. However, AXAF will cost over a billion dollars and EOS' estimated cost is tens-of-billions of dollars. Like the Space Station, there is great concern about the performance capabilities being overstated, about whether there is too much technological and management risk, and whether the requested budgets are deliverable. With this level of concern, the Executive Office and the Congress are beginning to explore new oversight mechanisms that could possibly limit the influence of these *de facto* incentives.

For AXAF, the Congress has chosen a pay-as-you-go approach wherein further funding is specifically dependent on the satisfactory accomplishment of technological milestones. For EOS, the Executive Office recently requested the National Research Council (NRC) to review EOS scientific objectives. Based on the NRC scientific conclusions, an external engineering review is at this writing being initiated to review NASA's current plan for flying EOS remote sensing instruments, as well as other credible alternatives. These external reviews may provide a fresh new perspective from a community better insulated from the forces of the *de facto* incentives.

Small unmanned projects. There continues to be little incentive to support boldly the Explorers, suborbital programs, and other small unmanned science projects. This is largely due to the perceptions associated with the large vs. small *de facto* incentive mentioned earlier. NASA program managers make most of the decisions at this level with input from NASA senior officials. The Executive Office and the Congress will routinely fund these efforts if the fiscal environment permits. There is usually little analytical or accountability review of these programs because of the respect for the merit review process.

Ironically, these programs offer extremely important and far reaching research and training opportunities, involve the fewest counter-productive *de facto* incentives, and will likely have the least need for external accountability. However, these programs also tend to be level-of-effort and a source for budget cuts because the constituencies are unorganized and relatively silent.

Conclusions

Much of the planning and accountability of our nation's space program has been delegated to NASA over the last several decades. Given the many NASA successes, this trust is not without a degree of reason and merit. It is also true that all parties (i.e., senior NASA, Executive Branch, and Congressional officials; the science community; contractors; etc.) have been responsible for the perpetuation of the *de facto* incentives and lack of accountability. Instead of a long-term, affordable human exploration program whose foundation is traceable to Apollo technology and a broad range of low-cost space infrastructure, we have the very costly and complex Space Shuttle and Space Station, and no appreciable human exploration program.

For the most part, if a problem occurs, the likely outcome is to relieve the program manager and pay the bill. This is not to say that the U.S. taxpayer has not gotten a significant return on our space investments. Operational needs have been met, there has been significant technological advancement and the creation of new knowledge, generations of first-rate scientists and engineers have been trained, and prestigious visions have been accomplished. However, the return on investment could have been even larger with better-managed programs. Little has been done to change the structure of the decision making process, incentives, or accountability mechanisms.

Nevertheless, the increasingly constrained fiscal environment and recent NASA problems (i.e., Shuttle, Station, Hubble, etc.) may be challenging this relationship and providing an opportunity to re-examine the decision making incentives that have dominated the space program in the recent past. In particular, there is a compelling need to greatly increase the incentives to fund small unmanned programs and less costly large unmanned and manned programs. In light of the repeated Space Station rescoping exercises, imagine the positive impact that even a small portion of the roughly $600 million spent on Station definition could have had on the science and technology research communities. One can easily see the tremendous opportunity costs of initiating overly ambitious or unrealistic space programs.

Some Thoughts on Possible Recommendations

A commitment to reduce the influence of some of these *de facto* program motives and develop a more long-term, integrated planning process could significantly improve our nation's space program. This means that all involved parties must further challenge the cost, complexity, implementation approach, and the scope of projects in the context of the nation's available resources. Projects need to planned relative to a realistic budget and astronauts should be included only where justified.

One possible approach to a new planning process would be to adopt some of the more open planning mechanisms developed by recent Federal interagency research efforts (e.g., U.S. Global Change Research Program, the High Performance Computing and

Communications Initiative, and the Math and Science Education Initiative). Each of these interagency research efforts have produced program and budget options based on a set of program priorities and evaluation criteria. Decisionmakers have made funding and policy choices based on these options.

A multi-agency effort could develop several options for a variety of long-term, visionary space programs. These options do not necessarily need to be developed by NASA, although NASA should be an integral player. They do need to be undertaken at the Cabinet level to ensure, from the beginning, that a formal and broad commitment has been made to undertake the program and the needed resources have been weighed against other national needs.

The plans should be reviewed widely to understand fully the various motives associated with each option. Criteria need to be developed to evaluate the various options, including technical readiness, relevance to the long-term objectives and program priorities, maturity of planning and cost estimates, and other pertinent evaluation measures. A method of evaluating program progress is also critical. Milestones could be set, not unlike the Congressional funding milestones established for the AXAF program.

Of course, this process does not need to be restricted to the development of an overall space agenda. An analogous mechanism can be established to develop project-specific alternatives, once the overall space agenda is agreed upon. Again, the lack of technically competent alternatives has handicapped the current NASA oversight process. In an era of $281 billion Federal deficits, to be reduced to zero over the next five years, the availability of lower cost alternatives must be the rule, not the exception.

Chapter 7
THE ROLE OF INCENTIVES AND ACCOUNTABILITY IN INDUSTRY AND GOVERNMENT

Angelo Guastaferro

FOR MANY ARE CALLED BUT FEW ARE CHOSEN.
—MATTHEW XXII:14

Introduction

Fundamentally, the civil space program is a very focused research and development activity. Because such activity is inherently unpredictable, the past thirty years have seen a government/industry-business relationship evolve around contracts in which legitimate costs of performing the work are reimbursed and the company's profit is in the form of an "award fee" based on performance. When a company begins a decade-long development project, it cannot tell what problems it may run into or what changes the government may demand; therefore, its costs can only be as well known as are the technical and programmatic requirements. Government and industry enter into a contractual agreement on estimated cost at the beginning of the program. Both parties agree that the award fee structure will be based on that negotiated cost. The fee or profit for a space program has been the element by which government has exercised the rewards and punishment system for the performing contractors. Over the past thirty years, the government has shifted from a straight "fixed fee" above estimated cost to an "award fee" system, which allows the government management team the opportunity to subjectively and objectively determine fee based on performance.

This paper will examine how program decisions and ultimate

program performance are driven by the rewards system. It will also explore what beneficial changes could be made in policies and to contractual implementation between industry and government. In addition, it will show how to cost-effectively improve shared program accountability and performance by better utilization of the reward system.

Monopsony Relationship

The business of civil space is, essentially, a monopsony relationship; i.e., there are many suppliers for one customer. This creates an environment in which the sole customer, NASA, can create demands on the contractor that force significant pre-contractual investment and aggressive pricing in a cost-reimbursable environment. Many qualified competitors respond to the one customer with an aggressive, positive, and optimistic proposal. This results in an industry tendency towards raised technological and programmatic expectations coupled with an aggressive commitment to schedule and cost. The competitive pressures within the federal budget for domestic programs and within NASA for doing more with less forces NASA to plan on more development programs than the agency can realistically accomplish. Reduction in the NASA budget request by Congress, which is now coming to be a regular annual event, causes the optimistic planning to result in an overall reduction in technical capability delivered at higher cost. This, ultimately, leads to legislative and public criticism of the project to the detriment of both NASA and the industry.

The competitive environment and the established NASA standards for adequate profit (i.e. fee) demand a pricing structure which yields profit margins below market expectations. The civil space aerospace industry gets a considerable amount of its new business investment supported by the federal government. However, the aerospace industry will provide considerable cost sharing from profit dollars to gain the competitive edge on large scale programs. (That is, a company will invest its own funds up-front, to win a contract, as is described below.)

A majority of aerospace firms are publicly held corporations where stockholders expect a reasonable return on their investment. The stock market is a strong indicator of the investing public's view of how the federal government allows for profit. An examination of the stock performance of the aerospace industry strongly indicates that investors do not consider aerospace stock a sound investment. This is especially true for companies with a significant part of their business engaged in government contracts. You may ask at this point, if doing business with NASA is so marginal, why are there still many sellers in the market place? That is a good question — I would like to spend the next few paragraphs addressing these motivational issues and how they may be used in future industry/NASA contractual relationships to achieve performance and accountability worthy of the space enterprise.

The NASA Decision Maker

NASA conducts its programs and projects in a phased procurement approach. The phases are divided chronologically and represent a continuing maturation of the endeavor. The program starts with a "phase A" period. NASA and industry conceptually look at a major program and conduct studies to determine the best set of architectures to accomplish the mission. It is typical during this phase of the program to involve many contractors at very little cost to the government. Phase A is normally accomplished in less than a year with resulting studies available to the entire government/industry community.

The next step, "phase B", usually involves two competing aerospace contractors. During phase B, the task is to identify the best set of functional requirements to provide the initial standard for a preliminary design. This phase is nominally conducted within a fixed-priced environment. The competing contractors will invest their own funds at this time to gain the competitive advantage leading to "phase C/D" which is always performed by the selected contractor for the final design, development, test and manufacture of the hardware and software needed to accomplish the mission.

The advocacy of a major program within NASA is a long and difficult process, which is described in the following paragraphs.

NASA is a federal agency organized to accomplish research, development, and operations in aeronautics and space. Institutionally, NASA has a strong central headquarters with several strong field centers. The headquarters' mission is to manage all programs and it has significant external responsibilities. It deals with the executive offices of the President, the Congress, the press, the scientific community and provides executive oversight to the many programs conducted at the several field centers. The NASA field centers are the backbone of the agency. These centers contain the technical personnel and research facilities to accomplish the day to day business of the agency. Each center maintains a variety of technical capabilities and strengths. They also manage the contracts with the aerospace industry during all program phases. The decision-making process leading to a major space program has its beginning at a field center.

The program must be justified through a process that involves competing aerospace contractors during the conceptual and program definition phases (phases A and B). At the same time, a strong program advocacy is undertaken by the NASA project management team to convince the field center director, the NASA headquarters program office, the program associate administrator and, eventually, the comptroller and NASA administrator that the program is ready for design and development (phase C/D). This is a long process that may take up to a decade. As an example, the Space Infrared Telescope Facility (SIRTF) has been in the advocacy phase since 1972 without proceeding to phase C/D. This bottoms-up approach to decision making provides an environment for the NASA program advocates to become very optimistic relative to the cost, schedule and performance of a particular program. This fact was recognized in a NASA study completed and reported in testimony to Congress in early 1980.[1] Under the direction of the then Director of the NASA Langley Research Center, Don Hearth, this study recommended that NASA pursue a non-advocate review prior to the initiation of the design/development (phase C/D) of a project. The non-advocate review, as developed, would allow a group of qualified NASA specialists not involved with the program the opportunity to review a proposed new development program for technical maturity, cost credibility and readiness to proceed. This practice has been accepted and in broad use since that time, significantly improving the NASA decision-making process. In fact, on a recent program to evaluate the readiness of the NASA/industry team to undertake the Assured Crew Return Vehicle (ACRV) contract start in FY92, the NASA Non-Advocate Review (NAR) team conducted detailed reviews at the NASA development center and the two competing contractor facilities. The NAR team reported their findings to the NASA administrator prior to completion of the FY92 budget process in the summer of 1990. Preliminary reports indicate that the NAR will recommend that NASA delay the program phase C/D start until 1993. The NAR also recommended adequate funding in FY92 for an extended phase B. This extension of phase B in fiscal year 1992 would allow the program to take several steps in order to greatly reduce schedule and budget risk when development (phase C/D) begins in fiscal year 1993. These steps would include:

- starting the process of procuring critical components whose manufacture involves long lead times;
- completing the development of enabling technologies; and
- initiating studies necessary to make "trades" between various competing approaches to solving technical and operational problems for the 1993 start.

Each fiscal year as part of the annual budget cycle, NASA rejects approximately ten program starts for every one approved. This high percentage of rejections tends to encourage the NASA manager to remain in the queue and continue with low-level funding to prepare for the next annual opportunity to start a program. This in turn encourages the competing contractors to continue contractor-provided funding and advocacy of the project.

Another of the recommendations of the Hearth Study was for NASA to invest sufficient early money in a program prior to phase C/D start. The existence of more good programs than NASA or the country can afford drives NASA to limit its early developmental funding on programs in the queue with the expectation that the competing contractors will make up the investment difference. The government procurement regulations have allowed aerospace contractors to invest in future programs by approving a small amount of discretionary money each fiscal year for the purpose of proposing on offerings from the government. This money is designated as "bid and proposal" (B/P) funds and is covered as an allowed expense of doing business. The same is true for investments in advanced technologies for the next generation of programs. This resource is identified as independent research and development (IRAD) funds. However, government procurement regulations have made it difficult to utilize B/P or IRAD money as a way of augmenting the government funds assigned to a phase A or B contract. This forces competing contractors into a choice of investing profit dollars or dropping out of the competition. I believe the government must make it legal for companies to utilize B/P and IRAD dollars as a way of augmenting the government investment during the phase A or B part of the program. If that is not possible, NASA must reduce the number of programs in phase A and B so that adequate resources can be applied to the surviving programs by both NASA and the contractors.

Finally, a negative aspect of a monopsony business environment is that outstanding contractor performance is rewarded with less profit. Both the NASA and the contractors have accepted this marginal condition. The space station prime contracts have been established with a maximum fee of 10% of costs for superior or 100% performance. In addition, that fee includes any cost allowances authorized by federal regulation (CAS414) as compensation for money borrowed for fixed assets in support of government work. That is, NASA considers costs allowed under CAS414 as another form of fee and reduces any awarded fee by the amount of the cost of money under the CAS414 authority. This rule can reduce the maximum fee of 10% to 7%. NASA should find a way to demonstrate a fee structure for outstanding performance closer to the 15% ceiling authorized by public law. However, the system should be designed for significantly higher reward for superior work and lower fee for marginal performance.

Industry Participation

The aerospace industry evolved from the national development of aeronautics. Technologies that took airplane development to higher and faster levels through World War II and into the supersonic era became the breeding ground for the emerging space programs of the sixties. With the maturation of the space program it was natural for the aerospace industry to continue to expand into the space technology arena. It was clear that investments in the enabling and enhancing technologies for space would be the path to successful participation in large scale civil space programs. Further, it became obvious that these investments would also provide a strong base for participation in the military and national security programs. Accordingly, over the past thirty years, the United States has developed a strong military and national security space program. This has led in turn to a strong technological base to support our civil space program. The application of hardware and software developments to the civil space program has been significant and will continue in the years ahead.

The aerospace industry understands the synergism between the military and civil space programs and how that synergism helps create a technology base that is beneficial to their overall aerospace business posture. In addition, one of the key resources of any aerospace contractor is its technical staff. The opportunity to work on high technology civil space programs helps attract and retain the best technical minds in the country. This leads to a highly motivated staff that makes an industry compete strongly to gain major civil space programs. Aerospace contractors maintain an aggressive advocacy role towards the civil space program. The long range survival of an aerospace company is directly related to the active and formative participation in programs that match their technological capabilities.

During the mission advocacy process, a very special relationship is established between the sponsoring NASA center and the competing phase A/B contractors. In order to demonstrate commitment, the contractors become eager agents in optimistic advocacy. The selling of the program becomes second to technology readiness. Again, the fear of being bogged down in the queue of projects ("many are called but few are chosen") drives the system toward significant contractor investment during the early conceptual phases. The longer it takes for final program development approval, the greater the investment and the more aggressive the advocacy on the part of government and industry.

Every contractor knows that winning the contract for a major program provides an opportunity for participation and funding for over a decade. This is another driver that makes for keen, aggressive competition and staying power during the phase A/B part of the program. If NASA could reduce the number of programs in the queue by identification of those programs that stand a strong chance of moving to the development phase in the future, they would reduce the number of programs to be supported by both government and industry. This would provide additional funding and better utilization of resources against the surviving phase A/B programs. Technical and programmatic planning during the phase A/B of the program will improve dramatically leading to a more accurate prediction of actual cost and performance.

Congressional Oversight

The Congress is elected by the public to make the laws which provide the authority and resource appropriations for the conduct of national programs. The Congress also has responsibility to oversee program performance. The process whereby NASA obtains its annual appropriations calls for NASA to defend the annual budget of the President as well as establish an early benchmark as to the cost, schedule and technical performance of each program entering phase C/D. The process calls for NASA to establish this early benchmark a year prior to receiving funding. This early marking coupled with aggressive advocacy and limited phase A/B government funding has been a major cause of misunderstanding of changes in program requirements and cost growth. Further, there has been a trend in

Congress to expand the quality and quantity of its staff. This leads to a greater visibility and access into NASA with a corresponding increase in programmatic micro-management. It is interesting to note how the last few appropriations bills for the NASA budget have drifted from legislative to executive language. Limiting funding authority and specifying when elements of a program should be undertaken is creating feelings of mistrust and contempt that can be detrimental to both space policy and program implementation.

The civil space program is in need of a funding profile that matches the national commitment. President Bush has provided, through the National Space Council, a clear set of goals for our civil space program including the Mission to Planet Earth, Space Science, and the Space Exploration Initiative. Congressional committee staff should concentrate on program understanding, fiscal needs and performance oversight. They should continue to approve the President's established national goals and to provide policy and funding for the effective implementation of those goals. Program implementation should be left to NASA with the Congress providing oversight on behalf of the people of the United States. Congress should seek the advice of the scientific community.

Research and development programs in support of civil space are high risk ventures requiring technical understanding and political support. The Congress must continue to demonstrate total support of the civil space program. They must be willing to sustain their support during program successes and be constructive critics during periods of difficulty. I believe the support the Congress showed after the Challenger accident is highly illustrative of what makes the American civil space program great. Congress demonstrated their sympathy for the families of the crew, assisted in helping identify the cause of the failure, and finally supported NASA during a very stressful period of technical and programmatic recovery.

Programmatic Impact

In retrospect, the achievements in civil space in the past thirty-two years have been a testimonial to the strong technological base of our government, industry and university system. Together we have accomplished much and have established aggressive planning to take us into the next century. This paper attempts to highlight a problem that is a product of the evolution of the NASA and industry relationship that has been at the forefront of the growth of civil space as a business. I believe that the environment surrounding the civil space program over the first thirty years has been influenced by the dilemma of having more ideas and programs than the federal budget could support coupled with a significant number of early successes. These successes provided a point of confidence that led to the creation of a trend towards fewer larger programs at the expense of a large number of smaller projects. The many suppliers-one buyer situation created a business environment that fostered an attitude in the aerospace industry that sustained more programs than the budget could support. This created a situation of an unlimited set of things to do against an insufficient budget.

I believe those forces will still exist in the years ahead but with less influence on the results. Unless some changes are made, some aerospace contractors will leave the competitive aspects of the business. If NASA intends to increase its budget to $30 billion by the year 2000, it will be necessary to look at changing the incentives for entry by increasing the funding for the program definition phase A/B. If this is not possible, NASA and industry must find ways for the industry to leverage B/P and IRAD resources toward underfunded phase A/B contracts. NASA must continue with cost reimbursable contracts with equally strong rewards and penalties. The government must be willing to reward the strong and penalize the weak. The space enterprise and the country need a government/industry team fully accountable for the success and failure of each program.

Recommendations

Both government and industry and their cooperative efforts are critical to the ultimate success of the U.S. Civil Space program. In an effort to improve the incentives and accountability in the performance of future space programs, it is recommended that the following be considered:

1. NASA should assume a greater share of pre-development funding.
2. NASA should reduce the number of underfunded programs and studies in phase A/B in order to provide adequate funding for approved programs.
3. NASA should continue the non-advocacy review process.
4. Congress should authorize multi-year funding for large programs.

5. Congressional staffers should provide better understanding of the risk associated with space exploration and get involved in the technical as well as political aspects of a program. (Become an informed critic.)
6. Industry, working with NASA, should find a way to leverage B/P and IRAD resources against underfunded pre-development programs.
7. Industry must work harder to put the best and brightest workers on programs to achieve superior performance.
8. Industry must demonstrate that they are willing to provide their best people and facilities towards the civil space program for the benefit of —

The Nation
The Public
The NASA
The Congress
The User Community
Their Employees
Their Stockholders.

Notes

1. D. P. Hearth "Statement", in U.S. House of Representatives, Committee on Science and Technology, Hearings, NASA Management and Procurement Procedures and Practices, 97th Congress, 1st Session, June 24, 25, 1981, No. 16, USGPO Washington, 1981. p. 11FF.

Chapter 8
POLICY ISSUES PERTAINING TO THE SPACE EXPLORATION INITIATIVE

Maxime A. Faget

HUMPTY DUMPTY SAT ON THE WALL,
HUMPTY DUMPTY HAD A GREAT FALL.
ALL THE KING'S HORSES, AND ALL THE KING'S MEN
COULDN'T PUT POOR HUMPTY TOGETHER AGAIN.

In October of 1957, the Russians launched the first artificial satellite to orbit the earth. This event, largely unexpected by the American public, the press, and the Federal Government, resulted in a strong response to catch-up and win the "space-race." The general approach was to supplement and convert some of the current effort devoted to developing short and long range ballistic missiles to building launch vehicles and spacecraft. The National Space Act created the NASA out of the NACA (National Advisory Committee for Aeronautics). Although NASA was not officially launched until October 1958, a great deal of preliminary planning and implementation of space programs was accomplished in the interim. This included drafting working agreements with the Army, Navy and Air Force which had the launch vehicles, tracking ranges, recovery forces and other supporting activity that the civilian space agency would need.

Today NASA is at a critical time in its existence. Its programs are being strongly criticized, particularly the Shuttle and Space Station – its high profile programs. Conditions could not be worse for inaugurating major new programs such as Earth Observing System and the Space Exploration Initiative, which would take us back to the Moon and to Mars.

Can an agency that landed men on the moon and explored the

solar system with robot spacecraft regain the public confidence? Or has it lost its ability to do great things? First it may be useful to understand what has happened and consider some possible causes. Perhaps from such a review possible changes can be identified that could reinvigorate the space program and reestablish NASA's well-deserved positive image.

During its 32 year history, NASA has accomplished a great many dramatic and exciting feats and has been regarded with a great deal of pride. NASA's most visible programs and, with a few exceptions, the ones which have been the most strongly admired as well as criticized, have involved manned flight. These are the programs most directly related to the Space Exploration Initiative. They are also the ones in which I, during my career as a NASA employee, have been directly and intimately involved. During this period, the political environment as well as NASA's organizational culture and management philosophy have experienced significant evolutionary changes.

The U. S. manned space program got its start with the Mercury Program, which was authorized in October 1958. The program start was preceded by a number of primitive *ad hoc* studies by industry, the Army, Navy and Air Force and by the Ames and Langley Laboratories of NACA. A team of less than a dozen Langley employees prepared preliminary specifications for Mercury during the summer of 1958. Shortly after the institution of NASA, the first NASA Administrator, Dr. T. Keith Glennan, created the Space Task Group to execute the Mercury program. Specifications were finalized and put out for bid in early November. Four responsive industry bids were received in December. Early in January 1959, McDonnell Aircraft Corporation was chosen and started work under an informal letter-of-intent contract by the end of that month.

Although McDonnell was under contract to develop and manufacture the Mercury spacecraft, the NASA employees in the Space Task Group using their own and other NASA resources from the Langley and Lewis labs performed a great deal of supporting tests and development activity that significantly contributed to the pace and success of the program.

Compared to current programs, Mercury was carried out at an amazing rate. McDonnell was under contract to build the Mercury spacecraft slightly over 3 months after the program was authorized.

Program History

Mercury
Preliminary Work Started	November 1957
Definitive Concept Established	August 1958
Program Authorized	October 1958
Contract Awarded	January 1959
First Significant Full Scale Test	September 1959
First Manned Flight	May 1961
Mission Achievement	February 1962
Program Complete	May 1963

Gemini
Preliminary Work Started	February 1961
Definitive Concept Established	July 1961
Program Authorized	December 1961
Contract Awarded	December 1961
First Significant Full Scale Test	April 1964
First Manned Flight	March 1965
Mission Achievement	July 1966
Program Complete	November 1966

Apollo (Command and Service Module)
Preliminary Work Started	March 1960
Definitive Concept Established	September 1960
Program Authorized	May 1961
Contract Awarded	January 1962
First Significant Full Scale Test	May 1964
First Manned Flight	October 1968
Mission Achievement	July 1969
Program Complete	December 1972

Apollo (Lunar Module)
Preliminary Work Started	January 1961
Definitive Concept Established	December 1961
Program Authorized	January 1962
Contract Awarded	December 1962
First Significant Full Scale Test	March 1969
First Manned Flight	March 1969
Mission Achievement	July 1969
Program Complete	December 1972

Space Shuttle
Preliminary Work Started	March 1969
Definitive Concept Established	February 1972
Program Authorized	March 1972
Contract Awarded	December 1972
First Significant Full Scale Test	July 1976
First Orbital Flight Test	April 1981
First Operational Flight (Mission Achievement)	November 1982

Space Station
Definitive Concept Established	August 1981
Program Authorized	January 1984
Contract Awarded	March 1985
First Component Deployed In Orbit	*Mid-1995
First Man-Tended Mission	*Mid-1996
Mission Achievement	*Mid-1998

* Scheduled Dates - Subject to Continual Revision

They delivered Al Shepard's spacecraft 22 months after a letter contract was signed. Shepard made the first U. S. manned suborbital flight 27 months after contract go ahead. Similarly, John Glenn orbited the earth 3 times in March 1962 — the first U. S. manned orbital flight — in a spacecraft delivered in July of 1961. Prior to Glenn's historic flight, 21 full scale flight tests had been made including 2 suborbital flights with astronauts and 4 flights (including one orbital) with primates. While all of the life-carrying flights were successful, there were 4 unsuccessful flights where either the launch vehicle or the spacecraft malfunctioned. The program was in fact carried out with the expectation of some failures since weaknesses were to be found and eliminated by flight testing.

The most dramatic example of the boldness and pace of the program was a full scale reentry test made using an Atlas "C" launch vehicle. In November 1958, the Space Task Group obtained the use of 2 Atlas "C" vehicles for flight tests in 1959. The Atlas would, of course, be able to toss a full scale test article into a trajectory that would simulate a reentry from orbit. Planning for the full scale test started immediately, but the final aerodynamic shape was not determined until the winning aerospace contractor (McDonnell) was chosen in January. The first flight was scheduled for August 1959, but slipped 1 month and was made in September. This flight verified the heating effects and stability characteristics of the Mercury configuration as well as the adequacy of heat protection materials used. It also included a full scale parachute and a recovery system. The test was more than successful since a malfunction of the Atlas launch vehicle aggravated the heating environment so that the test article actually survived a more rigorous environment than expected. Consequently, we canceled the use of the second Atlas set aside for this program.

Gemini and Apollo were also carried out at a much more rapid rate than present programs. Neither program was preceded by a long period of deliberation, requirement reviews or design studies. Gemini was started quite informally by simply negotiating with McDonnell (the Mercury contractor) for its development and construction without competition from other sources. It might seem to have been unfair to the other aerospace contractors, but it was, in our opinion, proper and least costly. And, most importantly it saved valuable time. Gemini provided the fledgling agency early operational experience needed for the more complex Apollo space flight missions. In fact, this greatly reduced the cost and program time needed to land men on the moon.

Apollo was announced by President Kennedy in May 1961, shortly after we launched Al Shepard on his short suborbital flight aboard a Mercury spacecraft. Prior to Kennedy's lunar proclamation, NASA had been planning to build a manned spacecraft that would be capable of lunar fly-by missions with lunar orbit as a follow-on program. Nevertheless, a lunar landing was made in just slightly more than 8 years from the time of Kennedy's announcement.

Compared to Mercury, Gemini and Apollo, NASA has progressed at a much slower pace on the Shuttle and the Space Station. This is illustrated on the chart which depicts the history of these programs from formal go-ahead to mission achievement. Also shown is the preliminary time that was spent in studies, etc. that took place prior to "go-ahead." For the purpose of this chart, mission achievement for Mercury was manned orbital flight; for Gemini, it was spacewalk, rendezvous and docking; for Apollo, it was lunar landing; for the Space Shuttle, it was first operational flight (STS-5); and for the Space Station it is considered to be when permanent occupancy is planned to occur. Program completion for Apollo is considered to be splash-down after the last lunar landing (Apollo 17) however, some residual Apollo hardware was subsequently used in the Skylab and the Apollo-Soyuz Test Project. Since preliminary work on a space station was started and terminated many times throughout NASA history, designation of a time for commencement of studies is inappropriate and consequently not shown although studies leading to the present design began in 1981. The table provides dates for the events reflected on the chart.

There are many reasons to believe that programs should be executed in the shortest possible time. From a simple economic standpoint, the financial, physical and human resources invested in a program have a time value which usually return little or no benefits until after successful completion of the program. Furthermore, technology is constantly changing and the product may lose some value through partial obsolescence. Freeman Dyson provides an eloquent discussion of this point in one of his Gifford Lectures.[1] Chapter 8 recounts his lecture called "Quick is Beautiful." While the whole chapter is an excellent treatise on the subject, as well as good entertainment, the following paragraph presents the meat of the message:

> Judging by the experience of the last fifty years, it seems that major changes come roughly once in a decade. In this situation it makes an enormous difference whether we are able to react to change in three years or in twelve.

> An industry which is able to react in three years will find the game stimulating and enjoyable, and the people who do the work will experience the pleasant sensation of being able to cope. An industry which takes twelve years to react will be perpetually too late, and the people running the industry will experience sensations of paralysis and demoralization. It seems that the critical time for reaction is about five years. If you can react within five years, with a bit of luck you are in good shape. If you take longer than five years, with a bit of bad luck you are in bad trouble.

Mercury, Gemini and Apollo were carried out at the fastest practical pace. The Shuttle and Space Station programs have been partially paced by a metering of financial resources through the annual budgeting process, but there are clearly many other factors involved. When a program continues to be extended, it soon becomes vulnerable to cancellation. The continued erosion in support for the Space Station, if not corrected, will also have a devastating impact on the proposed Space Exploration Initiative. Consequently, it would seem an urgent matter to examine all possible hindrances to the timely execution of major NASA programs and, to the extent appropriate, eliminate them. Future major programs should, if possible, be broken down into an overlapping series of quickly executed incremental programs, each of which can be justified on its own merit. For instance, instead of the current space station program an incremental program could start with a much smaller and simpler station that would require only 2 or perhaps 3 shuttle launches prior to productive occupancy. Occupancy would only be practical while the shuttle is attached to the station. Kits in the shuttle cargo bay would extend the duration of the shuttle time in orbit and augment the station life support equipment. Subsequent programs could expand this capability either with adjunct modules or co-orbiting freeflyers. Such programs could be more easily managed since they would minimize the size of the System Engineering and Integration (SE&I) function.

A series of incremental programs are also much more trackable from the standpoint of funding exigencies. The desirable profile of a program funding curve is bell-shaped. Truncating this natural curve with a planned annual funding cap is known to increase costs. Cost overruns become particularly troublesome and expensive. An overlapping series of smaller programs is a practical method of staying within a funding ceiling. It also provides management better options in dealing with reduced ceilings or overruns. For instance, curtailing the ramp-up of one of several overlapping programs in its early stages can be much more economical than de-scoping or stretching a single major program that is already in full bloom.

In reviewing the rate of progress of early programs in contrast to present ones, I could not help but reflect on many changes in the NASA culture during its relatively short history. By culture, I mean the context, both formal and informal, in which the NASA staff conducts their business. I am convinced that the average NASA employee of today is every bit as intelligent, dedicated and motivated as were the founding cadre. Statistically, they are better educated. What has changed are the institutional drivers and values. A minor and possibly completely unrelated example is that monetary awards and medals are now passed out much more freely than in the early days. There are many other detectable changes in the NASA culture, some of which were internally fostered while others were externally imposed or otherwise resulted from external stimuli. The following is a partial list of cultural changes. It should be noted that NASA cultural changes have been reflected to some degree in correlated changes within the aerospace industry.

- Boldness has been replaced by caution in program execution.
    Contrast the recovery from the Apollo fire to that
    from the Challenger disaster.
- Program management authority has largely been moved to Headquarters.
    Yet the expertise and long-timers
    (corporate memory) remain at the centers.
- NASA has experienced a decline in public interest following the lunar landings.
    NASA has had to adjust and structure its
    programs to political pressure supporting
    parochial interests.
- Extensive use of support service contracting is employed by NASA to supplement the size of civil service staff.
    Technical monitoring of contractor effort is
    replacing direct technical participation for a large
    part of the civil service staff.
- The acquisition process has become much more formalized and ritualized.
    This is a result of well intended acquisition
    regulations aimed at preventing fraud and
    favoritism that at the same time waste a great

deal of money in trivial time-consuming activity and endless documentation.

All of these changes have had both good and bad effects to some degree. However, the cumulative effect of all these changes over the years is bound to have some relation to the present situation. Many of these changes are either irreversible or only partially reversible from the standpoint of practical implementation and acceptability.

It is clear that it is not possible to return to the "good old days" at NASA. However, our nation cannot afford to fumble its response to the challenge presented by the new frontier of space. We must continue with a strong and vigorous civilian space program. NASA, as our civilian space agency, must be supported and strengthened. Programs must be selected and constructed so that they can be managed at one of the NASA development centers and completed in less than five years. Multi-center programs requiring strong Headquarter's management, such as Shuttle and Space Station, take too long to complete and weaken NASA's most important assets--its field centers and their experienced personnel. Rather than move the most experienced program managers to NASA headquarters, managers of the major programs should be moved back to the field centers. Programs that require the resources of more than one center should be broken down into smaller logical sub-programs that can be *independently* carried out at a single field center with a minimum need for control at the Headquarters level. The Apollo lunar landing program was really carried out that way. The Saturn launch vehicles were managed at Marshall while the Apollo spacecraft were managed at Johnson. Vision, policy, allocation of resources, and evaluation of results are the only productive functions of any good headquarters organization.

Notes

1. Dyson, Freeman, J., *Infinite In All Directions*, Gifford Lectures (Harper and Rowe, New York, NY; 1988).

Chapter 9
LOW COST ACCESS TO SPACE FOR SMALL SCIENCE AND TECHNOLOGY

Paul J. Coleman, Jr.

SMALL IS BEAUTIFUL.
—E. F. SCHUMACHER

Introduction

Research in space science and the development of space-related technology are important activities at many of the nation's universities. As with all university-based research and development (R&D), the space-related component serves two purposes: the advance of knowledge and the education and training of graduate students. More generally, a dynamic civil space program is believed essential to both the technological strength and economic competitiveness of the nation. It can be a source of technological innovation and talented engineers. The space program attracts young people to technical and scientific fields and can provide a training ground for advanced students in these fields. At the same time, the space program is a source of national pride.

My perspective is that of a member of the university-based R&D community and an educator of graduate students in engineering and basic science. Today this community is suffering from a lack of access to space for "small" projects of scientific research and technological development, especially those small projects in which graduate students are engaged.

Consequently, it is time to change the way we execute the unmanned component of the space program. In particular we should implement a systematic evolution in the mix of unmanned space flight missions towards smaller, simpler missions. In parallel with this implementation we should change the way we procure the services and products for smaller space-flight missions, especially the launch vehicles. Specifically, we should commercialize the small-missions component of our space program.

Some History

In the early days of the space program university-based scientists and their graduate students provided a substantial fraction, probably a major fraction, of the scientific instruments that were carried on our spacecraft. The rest were provided by scientists from government laboratories and industrial concerns.

With the passage of time, launch-vehicle technology improved. The payload capacity of our launch vehicles grew dramatically and our ambitions for spacecraft-borne observations grew apace. The more capable launch vehicles cost more. However, there were economies of scale so the cost per pound to orbit generally decreased.

At the same time that more payload mass was becoming available, progress in other technologies facilitated more sophisticated spacecraft subsystems and scientific instrumentation. As a result, we were able to do a lot more with each pound of hardware that we put into space.

During the years of this technological evolution, the process of mission planning, approval, and implementation evolved somewhat, but by and large it may be described as follows. Space technologists develop a more capable launch vehicle or a better technical capability of some sort, e.g., guidance to a planet. Space scientists are challenged to come up with a scientific mission to take advantage of this improved capability. A committee of scientists is established to consider this challenge. The committee works with NASA engineers in an iterative process to define a specific mission in some detail. Formal approval of the is new mission (approval for a "new start") is then sought, and, if obtained, the program is implemented. Competitive proposals are solicited from the scientific community. Scientists propose either to provide an instrument for the spacecraft as the principal investigator for that instrument or to participate on a NASA-selected team of scientists in using a NASA-built instrument.

The combination of technological progress and this program-planning process has moved the space program steadily towards more ambitious missions. Because the more cost effective launch vehicles are very expensive, the spacecraft that we launch are designed to be as capable as we can make them. Thus, the evolution of the unmanned, scientific program has been in the direction of maximizing the capabilities of every pound of payload that is launched.

The net result is that our ambitions grow as the technology improves and the cost of a spaceflight mission increases despite the fact that cost effectiveness of our hardware increases. Here, there is a clear analogy with military systems. Today's technology would allow us to do yesterday's mission cheaper than we could do it then. But we do not settle for doing yesterday's mission. Today's mission is so much more challenging that it costs more despite the more capable technology. Increases in cost effectiveness are accompanied by increases in cost. This is very different from what we have seen in the computer industry. There the technology has facilitated simultaneous increases in capability and decreases in cost: Increases in cost effectiveness are accompanied by decreases in cost.

NASA is an R&D agency, and it is appropriate that the NASA space flight programs be driven by technological improvements that lead to better performance and safety. At the same time, space scientists are driven by scientific objectives to develop better ways to acquire the data they need. These complementary interests have combined to produce the remarkable space-flight missions of the past twenty years.

Furthermore, these ambitious missions are highly visible in the national scene. They excite supportive public interest and produce high-technology jobs both directly and indirectly. Consequently, they generate a strong political constituency for the program.

The Problems

On balance, the unmanned science program has been a great success. However, it is not without problems, and some of these problems affect small science.

Early on we were caught in an upward spiral of costs. The cost per mission grew as our technical capabilities increased. As this cost grew, it became increasingly important to avoid failures. More extensive quality assurance and reliability (QA&R) programs were developed and added still more to the cost of the mission. (In passing, I would point out that these efforts towards absolute reliability carry an additional increment in cost. The hardware, especially the electronic components, has to be of proven reliability. An unfortunate result is that we use obsolete electronics, which are much more expensive than up-to-date electronics.)

These larger, more expensive missions are highly visible on the national scene, a mixed blessing. Such visibility fosters a political constituency for the program, but also raises the political costs of technical failures. Thus, it exacerbates the pressures to spend money to prevent failures.

As time passed, we could afford fewer missions, so we had to be even more sophisticated in our use of the payload mass, which made the cost higher, so the QA&R procedures had to be even more rigorous, which made the cost even higher, and the spiral continued upward.

Unfortunately, as matters now stand, it takes five to ten years and more from the go-ahead to the lift off of the spacecraft for these larger projects, and a project often costs a billion dollars or more. These larger-scale missions in the unmanned program have become so complex and so expensive that we cannot afford to have one fail. Accordingly, we have set a goal of essentially zero failures. In such a situation only professionals, and only the best professionals at that, can participate in the design, fabrication, launch, and operation of these magnificent machines. They are properly and, in fact, necessarily the products of high-tech industrial concerns and national laboratories, not universities.

Unfortunately, the process, procedures, cost, and time associated with the larger unmanned space flight missions now typical of NASA's program militate in many ways against progress in small science and the participation of students. The university professor and his or her students, and even the junior scientists and engineers at the governmental and industrial organizations are not likely to be involved in the development, fabrication, and operation of these magnificent machines. Because so few missions are launched, an experimentalist cannot rely on the usual process of performing an experiment, learning from the experiment, performing a new experiment, etc. Furthermore, space-flight instrumentation for an experiment cannot be designed, built, launched, and operated during the tenure of the average student in graduate school.

In summary, the high cost of access to space has imposed especially severe limitations on the access to space for small projects of scientific research and technological development. This lack of access to space is slowing scientific research, stifling technological innovation, and reducing the hands-on training available to our graduate students in engineering and science. Furthermore, as I will discuss below, it is reducing the cost effectiveness of the space program.

A Solution for Small Science and Technology

We should implement systematic evolution in the mix of unmanned space flight missions; an evolution that moves this part of the program away from larger, relatively complex missions and toward smaller, simpler missions. In concert with this shift, we need to commercialize the procurement of the goods and services required for the small-missions component of the program. Such a dual-track effort would trigger the development of truly low-cost access to space.

Because NASA is an R&D agency it is appropriate that NASA space flight programs are driven by performance, reliability, and safety. The disadvantage of this combination of drivers is that the cost of access to space increases rapidly.

By contrast, what is needed for the small-missions complement of the program is a different combination of drivers, namely, cost, reliability, and safety. In this part of the program, we need conventional product engineering to provide hardware that is cheap, reliable, and safe. It seems to me that these drivers are appropriate for commercial endeavors. Thus we should commercialize the procurement of goods and services for the small-missions part of the space program.

Simply stated, the objective of this dual-track effort would be the capability to fabricate and launch small spacecraft, including earth satellites and lunar, planetary and interplanetary spacecraft, at costs comparable to today's cost for one of the larger sounding rocket payloads. The cost must be low enough to permit a relatively large number of launches; a number perhaps smaller than the number of sounding rocket launches, but certainly much greater than the number of space flight missions. Under this condition, spacecraft can be relatively simple and the cost of integrating numbers of complex instrument systems on a single spacecraft can be eliminated.

The cost must also be low enough to justify the elimination of many of the quality assurance and reliability functions that are not associated with safety. This specification is especially important because it will allow us to reverse the upward spiral of cost associated with increasing QA&R.

It follows that we would be able to do a substantial component of the work of the space program more cost effectively.

Small-Mission Programs at NASA and DARPA

In 1988, NASA implemented its program of Small-Class Explorer Missions. This program is a big step in the right direction. As a result, we now have an ongoing program upon which to build. (It is essential that individual projects of the magnitude required for a low-cost program should not require line-itemization, and therefore line-item approval, in the Federal budget. The Small-Explorer program includes this feature.)

Also, both DARPA and NASA have taken important steps towards cheaper small ELV's. DARPA has funded the development of a "small standard" launch vehicle called Taurus. The first stage of Taurus is the first stage of a small Peacekeeper missile. NASA has ordered a number of Orbital Sciences Corporations's aircraft-launched ELV's called Pegasus. Also, the DOD required that the medium launch vehicle (MLV), the Delta 2, be designed with commercial applications in mind. However, we have some distance to go in reducing the cost of ELV's.

From my perspective, the small missions under current consideration by NASA are simply smaller variations of large missions. There is too little innovation in the approach to design, fabrication, reliability, and quality assurance. There is too little use of new technology. Consequently, these small missions are still too expensive. NASA estimates something like $30 million for the cost of a small explorer mission.

Because the NASA-defined small missions are still too expensive, there is no component of the NASA program in which we can learn by trying and – inevitably – failing from time to time. There is no component of the program in which we can experiment with new ways to do things.

Similarly, as I just mentioned, the small ELV's listed above will probably be too expensive. Costs between $10 and $20 million plus per launch have been estimated for DARPA's Taurus and NASA's Pegasus ELV's. Consequently, our access to space will continue to be constrained, I believe unnecessarily, by the high costs of the business.

Why Now?

In the past, technological, institutional and bureaucratic obstacles have deterred the development of low cost access to space. Why should this development be easier now? First, I believe the technological obstacles no longer exist. Second, there are reasons to be optimistic that the others can now be overcome. In particular:

1. There is a growing concern about the technical problems that affect the reliability of the larger spacecraft that are required for the major space flight mission. NASA and ESA have had serious problems with the eight or nine most recent of their major spacecraft. Representative of this concern is the intense discussion that has been precipitated by the size, complexity, and cost of the spacecraft that NASA plans to use for the EOS-A mission. We may well have reached some kind of inherent limit as to the complexity of spacecraft that we can expect to work perfectly the first time they are put in operation. (Perhaps such spacecraft are at a threshold: They are at once both so tightly integrated and so complex that no one can foresee all possible failure modes − yet adding technology intended to increase reliability only increases interdependence, complexity, and the number of possible failures with a net decrease in reliability resulting.)
2. There is general recognition that graduate-student education and training in a number of space-related, engineering and scientific specialties could be better served by the unmanned component of the civil space program. The statement is frequently made that balloons, aircraft, and sounding rockets provide the only training grounds for experimentalists in space science. I suspect that a similar statement can be made relative to the hands-on training of engineers in space-related engineering specialties.
3. Most of the community of space scientists and engineers has until recently been on standby because of the problems at NASA that restrict access to space. Consequently, there is a strong, pent-up demand for small missions.
4. The space-related research and development community requires a range of opportunities to maintain stability that is essential to the well being of any academic research specialty. At the same time, there is the feeling that future budgets will

not support the number of currently planned missions that is necessary for this maintenance.
5. Low-cost access to space will provide a fertile field for technological innovation, and the Congress is aware of the importance of technological strength to economic well being in a modern, competitive world.
6. Several companies, mostly new starts, have undertaken to commercialize one or another of the services essential to the implementation of a truly low-cost, small-mission component of the civil space program. These new companies include talented, energetic people who have struggled mightily to overcome the obstacles to such commercialization.
7. The growing interest in space-flight projects in other countries suggests a significant potential for sharing such a program and its fixed costs with others in the international community. (Such a collaborative thrust is another element of NASA's program of Small-Class Explorer Missions.)

I stated above that the technological obstacles no longer exist. There is a host of activities, in both the public and private sectors, under which relevant systems, subsystems, components, software, and procedures are being developed. In general, I am convinced that:

1. Off-the-shelf electronics can go into space with minor or no modifications and can satisfy all but the most severe restraints on weight and power consumption. Lap-top computers have been carried on the Shuttle.
2. Low-cost launch vehicles can be developed. Above, I discussed the efforts of NASA and DARPA to develop a lower-cost small ELV. At the same time, several smaller companies such as Microsat and Amroc, in addition to Orbital Sciences mentioned above, are working on their own toward this objective. Alternatively, modified versions of existing sounding rockets or military boosters may provide the lower cost ELV's needed for small science. Among the candidates are Lockheed's Poseidon-based Lodestar, Boeing's Minuteman II-based vehicle, Space Corporation's Peacekeeper-based launch vehicle, and Orbital Sciences Peacekeeper-based Taurus, mentioned above.
3. For most experiments off-the-shelf memory can provide on-board storage sufficient to permit tracking, control, and data acquisition from a single station on the ground.

4. A "standard", low-cost earth satellite could have the capability to re-enter upon command, either active or programmed, so that these spacecraft would not add to the debris in space. The University of Colorado is studying a modified version of Biosat as one candidate for re-entry missions.
5. The capability to deliver payloads to the Space Station and other orbiting platforms by rendezvous with an orbital maneuvering vehicle could be readily developed on the scale that is relevant to small missions.
6. The capability to transfer to GEO could be developed.
7. Eventually, the capability to reach the moon and planets could be provided.

The feasibility of these last two extensions is suggested by the early results of a study at the Jet Propulsion Laboratory (JPL) of the "Lunar GAS Satellite." Specifically, JPL is working on a design of a 250 lb. spacecraft that would be launched from a Getaway special (GAS) canister on board a Space Shuttle. The spacecraft is designed to orbit the moon and serve as a platform for remote sensing instruments.

As mentioned earlier, I believe that the currently advanced state of space technology will permit us to effect this change in the mix with no reduction in our capabilities to do space science. That is to say, we can do a significant part of the work of the unmanned scientific program with smaller, simpler spacecraft, and, because of the current state of our technology, we can do this work with no decrease in the quality of the results.

Commercialization Defined

It seems that the procurement process used in the space program is inimical to low-cost access to space. In considering how to improve this process, I have found what seems to be a relevant analogy. Not many years ago, the digital computer was an item of special purpose instrumentation so expensive that only a few of the nation's research centers had one of any significant capability. Nearly all such machines were considered to be major research facilities, much like the larger accelerators and astronomical telescopes of today. They were designed and built under contract from a Federal agency and made available to researchers as "government-funded equipment" or GFE.

In other words, we got our computers then just the way we get our launches now.

Today we at the universities get our computers through a different procurement process. Although the funds for computers are still Federal funds, the procurement is not done directly by the Federal government. Rather, it is done by the universities, using funds usually derived from Federal agencies through contracts with the institution. Before this change in the procurement process, only a few companies served the market. Since the change, many companies now make substantial profits selling computational equipment and software to the university-based R&D community. At the same time, the cost per unit of capacity has fallen dramatically. Thus, normal market forces have acted to provide low-cost access to computational power. Today computers (of much greater capability) are essentially hardware utilities and they are purchased accordingly.

I claim that many of the tools and services used by space scientists and engineers are now, or rather should now be, in the utility category. I include in this category spacecraft; spacecraft systems such as power, communications, and attitude control; and launch vehicles; as well as a number of launch services.

Consequently, as a step toward developing a market, these tools and services should be purchased directly by the users, using funds from government grants and contracts as appropriate, rather than being purchased by the Federal government and provided to the users. In my judgement, this procurement-process alone would be a giant step in the reduction of the cost of access to space.

The keys to the ability of this approach to reduce costs are development of a credible marketplace for low-cost, space-related goods and services and a credible technological base from which to provide these goods and services. Before we changed the way we purchased computers, the need and the technological base were there, but the marketplace was not. With the change in the procurement process the marketplace became apparent to the commercial computer industry and the business took off. I believe that this same, commercially-driven process would work in the space industry.

Accordingly, we should commercialize the procurement process in the following way: We should change the procurement process so that university and industrial scientists under grants and contracts with the Federal government to perform some program of space research can use the Federal funds in their grants and contracts to purchase from commercial vendors the goods and services that will let the R&D

contractors, be they universities or non-profit companies or commercial entities, buy launch vehicles and services on an open market. (Universities do this now on spaceflight instruments, some costing tens of millions of dollars, so I believe they are in general perfectly capable of handling such procurements.)

The other track of this dual-track approach of course concerns this marketplace, which I have discussed above. NASA must implement the evolution towards smaller missions so that there can be enough transactions to define a market. As described above, they have taken a step in the right direction with the NASA-defined small missions. However, these projects are still far too expensive. Both the university-based R&D community and the commercial space launch industry must be convinced that these "commercialized" small missions will become an important part of the program.

Recapitulation

In summary:

1. A systematic evolution toward a low-cost, small-mission component in NASA's program would create a quantifiable market for the products and services that would be needed for these small missions.
2. The commercialization, as defined herein, of the procurement of space-flight-related products and services for this small-missions component would create a marketplace for the commercial providers of such products and services.
3. The combination of these two steps would galvanize the remarkably talented and dedicated group of space "entrepreneurs" in this country, a group of engineers and scientists from government, industry, and the universities who are already working to provide us with low-cost access to space. It would stimulate the development of truly low-cost access to space.
4. This would significantly expand access to space for the small science and technology community. It would increase the number of flight opportunities and improve the quality of the participation of those involved.

5. This expanded access would require no increase in the budget of the space program and would produce no decrease in the quality of the scientific results from the space program.

In short, these changes would make the civil space program a more effective partner in the education and training of advanced students in engineering and science, would make the program more cost effective, would create a new wave of innovation in space-related technology, and would spawn an exciting new component of private enterprise.

The obstacles to this process are those of policy and precedent, rather than budget. Nevertheless, the obstacles are formidable. An appropriate policy statement from the White House might be adequate to implement the process. However, at this writing, it would appear that legislation or at least a Congressional mandate is needed.

Part Three
PROGRAMS

In this part five chapters address specific space program areas and show how they are affected by the policy context. Thus this part takes a program perspective and looks at the effect of context on them, a view orthogonal to that of the previous part. The authors suggest policy changes that could be made to improve performance.

Two of the essays are on science programs. McCray and Stern look broadly at science programs and make some sweeping recommendations. Stern and Habegger look at one particular aspect of science programs, the transition to single mission programs. They were able to find quantitative data to apply to a space policy question, i.e. why and to what effect did the transition occur.

The third chapter in this part discusses NASA's Earth Observing System, known as EOS. It is a science program, but has a very clear practical mission and thus has some aspects in common with what NASA calls applications programs, e.g. communications satellite programs. The ambitious goals of EOS raise several questions in Webster's chapter.

Two chapters discuss manned programs. Brunner et al. discuss the cloudy future of the Space Station program. Pielke and Byerly compare the performance of the Space Shuttle program to the promises made in selling the program and find disparity.

A common thread running through all the chapters in this part is that successful execution of any kind of large, inflexible program will be difficult in today's policy context. Evolutionary approaches and small steps are likely to be more effective. If large scale goals are to be achieved, effort should be made to break them into small, manageable parts.

Chapter 10
NASA'S SPACE SCIENCE PROGRAM: THE VISION AND THE REALITY[1]

Richard McCray and S. Alan Stern

> "HE'S DREAMING NOW," SAID TWEEDLEDEE, "AND WHAT DO YOU THINK HE'S DREAMING ABOUT?"
> ALICE SAID, "NOBODY CAN GUESS THAT."
> "WHY, ABOUT *YOU*!" TWEEDLEDEE EXCLAIMED, CLAPPING HIS HANDS TRIUMPHANTLY. "AND IF HE LEFT OFF DREAMING ABOUT YOU, WHERE DO YOU SUPPOSE YOU'D BE?"
> "WHERE I AM NOW, OF COURSE," SAID ALICE.
> "NOT YOU!" TWEEDLEDEE RETORTED CONTEMPTUOUSLY. "YOU'D BE NOWHERE. WHY YOU'RE ONLY A SORT OF THING IN HIS DREAM!". . .
> "I KNOW THEY'RE TALKING NONSENSE," ALICE THOUGHT TO HERSELF, "AND IT'S FOOLISH TO CRY ABOUT IT." SO SHE BRUSHED AWAY HER TEARS, AND WENT ON, AS CHEERFULLY AS SHE COULD.
> —L. CARROLL

Introduction

Recently, the "Augustine Committee" released its *Report on the Future of the U.S. Space Program*[2] to the Administrator of NASA and the Vice President. One of the major recommendations of this far-reaching, visionary report is that NASA's space science program warrants highest priority within the agency, ahead of space stations, aerospace planes, and manned missions to the planets. This recommendation, if accepted, is good news indeed for space scientists, who have suffered a long hiatus of launch opportunities during the 1980s. In order to make the most of this opportunity, we should re-examine NASA's space science program to see where gains may be made in its productivity.

During the next five years, NASA's Office of Space Science and Applications (OSSA) plans to launch some 40 scientific payloads. These include a number of long-awaited, large missions, such as the Gamma Ray Observatory (GRO), the Upper Atmosphere Research Satellite (UARS), and the TOPEX oceanography mission.

Beyond that, OSSA, under the leadership of Dr. Lennard Fisk, NASA's Associate Administrator for Space Science and Applications, has developed an exciting 5-year Strategic Plan[3] for starting new projects that, if implemented, would maintain this pace throughout the 1990s with the launch of a major mission (of the GRO or Galileo class) every year or two, one or two Small Explorers (SMEX) each year, a moderate mission (*e.g.*, the Extreme Ultraviolet Explorer, Mars Observer) every two or three years, and a variety of other experiments on the Shuttle and on foreign missions.

According to the scenario envisioned in the OSSA strategic plan, space science is about to enter a renaissance that will even surpass the decade 1966 – 1975, when some 70 scientific satellites were successfully launched.

However, this scenario is under duress even as we write. The Hubble Space Telescope (HST), the second most expensive space science project in history (after the Viking Mars lander), was launched with a faulty mirror and a thermally excited vibration that deeply compromise its ability to obtain the sharp images and spectra of faint stars and galaxies for which it was designed. Although the HST will still be able to do frontier science at reduced efficiency, and although it probably can be repaired during a future Shuttle mission, the cost of this repair will delay other scientific missions and may raid small, individual investigator groups of development funds. Moreover, the main antenna of the Galileo mission has failed to deploy properly with the possible consequence that the spacecraft will not be able to achieve its primary scientific objective of sending back images of Jupiter and its moons.

These painful outcomes are to some degree inevitable consequences of the risk involved in space flight. However, they are symptomatic of a chronic fragility of programmatic planning that has caused severe problems in the US space science program for the past decade.

The problems began in the 1970s with NASA's vision of a space science program tied to a transportation system built around Shuttle and a variety of orbital transfer vehicles. Accordingly, it would be possible to launch and tend many large attached and free-flying payloads in a variety of Earth orbits and to launch heavy payloads from the Shuttle into geosynchronous and interplanetary orbits, all at unprecedented low cost. In such a heady atmosphere, many scientists neglected warnings about the dangers inherent in the complex interfaces between the space science program and the Shuttle Space Transportation System (STS). Likewise, administrators and politicians ignored warnings about placing too much demand on an experimental, manned Shuttle with too little fleet capacity.

With hindsight, one can see how this flawed vision led to the crisis of space science in the 1980s. The Space and Earth Science Advisory Committee (SESAC) report[4] entitled "The Crisis in Space and Earth Science" documented this, and the 1990 OSSA Strategic Plan incorporates many of the lessons that have been learned. However, as we describe below, the crisis of the 1980s resulted from a mutually reinforcing dynamic involving the scientific community, NASA, the OMB, and Congress that continues and that will likely again cause problems. In particular:

- Strong institutional and political forces favor big missions over small ones.
- Unrealistic budget planning removes resilience in the planning process, resulting in reduced productivity.
- A culture of technical risk-avoidance and programmatic risk-taking exists which is detrimental to an innovative and productive space science program.
- Pressures persist which encourage inappropriate linkages of scientific experiments to the manned program.

These difficulties are not insurmountable. Now that NASA is clearing the backlog of scientific missions delayed by the Space Shuttle, OSSA has a chance to carry out an exciting program of space science that will be much more robust and productive during the latter part of this decade and beyond. We believe that the way to achieve this goal is clear; indeed, many of the recommendations we discuss are not new[5] and some are already being implemented by OSSA. We urge that the scientific community, OMB, Congress, and OSSA work together to:

- Change the mix of new space science missions so that less than half of the OSSA budget is devoted to *large* (>$600 M) missions, and substantially greater funding becomes available for *small* ($30 - 150 M) and *moderate* ($200 - 500 M) missions.
- Invest a greater fraction of resources in the early development phase of missions, some of which may not be selected for completion. Further, OSSA should consider a wider range of mission alternatives when selecting a concept to achieve a desired scientific objective.
- Not make a commitment to build any space science mission until the mission is sufficiently well-defined that a *realistic and firm* budget (with adequate reserves) and schedule can be established.
- Once selected, space science missions should be built to this budget and schedule, *even if it becomes necessary to de-scope the scientific performance of the mission.* This will require that OSSA and the mission management team have full authority for the conduct of each mission, including choice of launch vehicle and operations strategy.
- Full authority and accountability for mission development should be vested in a tight management team led by the project manager and the principal scientific investigator.
- The scientific program should be coupled to the manned program only when there are compelling scientific, technical, or economic justifications.

These recommendations are based on a philosophy of reduced programmatic risk which we advance below. They are designed to enhance the resiliency of the US space science effort against internal and external perturbations (*e.g.*, cost over-runs, launch failures, budgetary constraints).

The SESAC Report and OSSA's Strategic Plan

An important and striking point made by the 1986 SESAC report is that the crisis in space science was not caused by the Challenger accident, but largely by a concentration of resources in a diminishing number of space missions of increasing size. The signs of crisis, evident long before the accident occurred, included: (1) the timely funding of planned missions was becoming increasingly unreliable; (2)

missions were suffering massive cost overruns and schedule stretch-outs; and (3) the consequent long delays in launching scientific payloads and the insecure prospects for the future were leading to the demoralization of scientists and erosion of expertise in disciplines in space science. The Challenger accident exacerbated these problems and brought the issues into sharper focus.

The major recommendations of the SESAC Report were that:

- The OSSA program should incorporate a diverse array of scientific activities, participants, and facilities. In particular, OSSA should strive for diversity in the scale of missions, a vigorous and stable research and analysis program, and involvement in substantial international collaborations.
- The scientific requirements of missions must be the dominant factor in the choice of the launch vehicle. This recommendation included a clear call for greater reliance on expendable launch vehicles (ELVs) for scientific payloads.
- A more rigorous planning process should be instituted in the OSSA program, including more realistic assessment of mission costs and resources and tighter control of costs and schedule. In particular, NASA should not make commitments to the definition and design (Phase B) of missions in the absence of convincing evidence that the missions can be started and completed on schedule, and that the financial resources will be available as required to support the proposed schedule.

Recent OSSA Strategic Plans have responded to many of the recommendations of the SESAC Report. The Plan is updated annually and establishes priorities for decisions to be made over a five-year period. It envisions a diverse scientific program, implemented according to a set of well-defined rules. The first rule is to complete the ongoing program, which today consists of several missions expected to be launched during the next few years, including some 14 specified free-flying satellites and several Spacelabs under development. The second rule is to initiate a major or moderate mission each year; the Plan lists seven such missions in order of priority. In addition, the Plan calls for: (1) the initiation of a program of small Explorers

(SMEX); (2) augmented support of the research and analysis base; and (3) development of scientific instruments for the Space Station.

The OSSA strategy calls for a renewed reliance on expendable launch vehicles, with the launch of roughly two *small* (Scout class), and two to four *intermediate* (Delta, Atlas/Centaur, Titan III) vehicles each year, and one *large* (Titan IV) vehicle every two to three years; it also plans for continued reliance on the Space Shuttle at the rate of four or five launches per year. The high launch rate implicit in this plan is necessary to remove the backlog of payloads awaiting launch since the Challenger accident, and to fly a series of Spacelab missions which have been under development for a decade. A lower rate will be maintained after the backlog is launched.

The Strategic Plan assumes that OSSA will be able to work with a budget that will remain at roughly its current 20% share of the total NASA budget, and that the total NASA budget will continue to grow throughout the 1990s with an average annual increment of roughly 10% above inflation. This would permit the start of one major or moderate mission per year for the next few years. According to the Plan, if the budget does not increase at the projected rate, the development of the major missions will be stretched out or delayed, but the moderate and small missions would proceed at the planned rate.

The OSSA Strategic Plan is a decisive response to many of the issues and recommendations raised in the SESAC Report. Will it work? When asked what he considered the greatest threat to the Strategic Plan, Dr.Fisk answered without hesitation: "The budget."[6]

Persistent Threats to Space Science

The productivity of NASA's Space Science program is threatened by: (1) the concentration of resources in large missions; (2) the culture of technical risk avoidance and programmatic risk taking; (3) non-resilient budget planning; (4) inappropriate linkage to the manned space flight program; and (5) an erosion of technical and management expertise within NASA and the space science community. These issues reinforce each other, as we now describe.

The Concentration of Resources in Large Missions

The SESAC Report clearly identifies the concentration of resources in missions of increasing size and complexity as a major factor contributing to the crisis in space science. The most obvious consequences were a decreasing launch rate and increasing lead times for scientific payloads. The average launch rate declined from 8.5 missions per year in the 1960s, to 4.7 per year in the 70s, to 2.3 per year in the pre-Challenger years 1980--1985. The time between starting a mission and its launch increased from 6 years in the 1970s to more than 13 years for those missions started in the mid-70s.

The tendency to concentrate resources in major missions has several consequences. Missions such as the HST, Galileo, and the Advanced X-Ray Astrophysics Facility (AXAF) each cost (including launch) $1-2 billion or more, compared to $200-500 M for moderate (*e.g.*, Explorer, Planetary Observer) missions and roughly $60-120 M for small (Small Explorer, Earth Probe) missions. This imbalance goes a long way toward explaining the decrease in launch rate: OSSA could afford to build and launch up to 5 moderate missions or 15 small missions for the cost of each major mission.

The interval between conception and launch of major missions has now grown to some 15-20 years. Consequently, the health of entire sub-disciplines (such as X-ray astronomy or lunar science) may rely on individual spacecraft separated in time by a substantial fraction of a professional career. Given the inherent risks involved in space flight, such a strategy introduces an unacceptable degree of programmatic risk into the space science program. One can easily imagine the setback that would be incurred by a large investigator community if a major mission such as the AXAF, TOPEX, or the Cassini Saturn mission were to fail completely on launch or before accomplishing its objectives. An alternative strategy involving a reduced dependence on such large missions would mitigate this programmatic risk and increase scientific vitality, as we discuss below.

Certainly, large missions can provide scientific advances that cannot be made with smaller missions. For example, a major force driving the size of the HST was the requirement for high angular resolution, which is proportional to the mirror diameter. Thus, given the large investment required for the mirror and accompanying large spacecraft bus, having four separate, powerful instruments that could be replaced and upgraded was very attractive to scientists. The penalties of such a strategy are now becoming apparent, however. One is the risk associated with concentrating a large fraction of the Nation's space science resources in a single mission, which can subject an entire scientific discipline to the potential for a technical single-point failure. Large missions also introduce losses in productivity. When many instruments share a common telescope, the scientific utility of each instrument is often reduced to a fraction equal to its share of the telescope time. Further, the whole telescope and spacecraft must be designed to the accuracy and pointing requirements of the most demanding instrument, and this very expensive capability is wasted on instruments with lesser requirements. Thus today some of Hubble's instruments are able to do good science using the degraded optics.

A simple hypothetical model will illustrate how it might be advantageous to break a large space telescope into two smaller ones. For space astronomy, the number of useful scientific observations that can be made in a given time is usually proportional to the light-gathering area, or D^2, where D is the diameter of the telescope. Experience in aerospace industry projects shows that the cost of a spacecraft is roughly proportional to its weight, which for a telescope, is roughly proportional to $D^{3.7}$.

Now, consider the cost/performance trade-off in the choice between building one 3-meter telescope with a full suite of instruments or two 2-meter telescopes, each equipped with half the total payload complement. Suppose the 3-meter telescope costs $1 billion. Then, according to the cost-weight relationship, each 2-meter telescope will cost $(2/3)^3$ times the 3-meter telescope, giving a price tag near $600 M for the pair. The savings may be even greater if similarities between missions can be exploited to achieve economies in production. Performance? The net area of the two 2-meter telescopes is almost the same (8/9) as that of the single 3-meter telescope. Thus, in this simple example, splitting the mission into two would yield a 41% cost savings from an 11% sacrifice in light gathering power. But even greater advantages to science accrue from the increased resilience to failure, more opportunities to train instrumental scientists, and the potential to improve the second telescope based on the experience gained from the first.

Of course, this model is oversimplified. In reality, an exercise in optimizing scientific productivity should also take into account other important performance characteristics, such as image resolution (which would be decreased in our example). Nevertheless, the example illustrates that there *are* opportunities to substantially increase the productivity and lower risk in NASA's space science program by shifting resources toward smaller missions. Unfortunately, however, as we discuss below, political realities militate against such rational choices.

It is almost dogma within NASA and the scientific community that as space science progresses in a given discipline, there is a natural evolution to missions of the HST or Galileo scale. Indeed, some space scientists claim that the data return per dollar is greater for a large mission than for a small mission. As the example above shows, however, this assertion is debatable. Moreover, recent missions such as Active Magnetospheric Particle Tracer Experiment (AMPTE) and the Cosmic Background Explorer (COBE) show that moderate missions can still deliver frontier science.

Even if big missions were more efficient than small and moderate missions in terms of the dollars spent for data generated, there would still be a question of whether that ratio is the only appropriate figure of merit. Scientific vitality is not measured simply in terms of quantity of data. Other considerations may be more important--such as the continuity and stability of a given discipline and the opportunities for training new scientists and instrumentalists. Likewise, large missions do not provide opportunities for bringing new inventions to fruition quickly. Indeed, it is often necessary to select instruments for large missions many years before launch, precluding the use of more advanced technologies that may be developed in the interim.

Moreover, as the HST experience shows,[8] the technical complexity and long lead times of large missions can lead to a diffusion of management authority and lack of continuity in technical management that increase the likelihood of cost overruns and the risk of technical failure.

We believe that OSSA and the scientific community are often driven to missions of ever-increasing complexity and cost by an interplay of several factors, such as: the desire by scientists to achieve the maximum scientific performance of each mission; the tendency by NASA to make optimistic projections of mission costs; the reluctance of both NASA and scientists in the community to trade off performance goals in order to preserve cost and/or schedule; the budgetary uncertainty resulting from the annual appropriations process; and the difficulties imposed by the Executive Branch and Congressional approval process which favors a small number of large missions over a larger, more resilient number of small missions. The interplay of these factors has caused the resources for space science to become concentrated in a diminishing number of missions, thereby reducing the scientific productivity of the program as a whole, even while attempting to maximize the performance of each individual mission. Such a system fosters programmatic risk and, ultimately, instances of keen disappointment.

This dynamic is institutionalized in the procedures for selecting new space science missions, which are selected and entered into OSSA's queue of planned new starts with the help of scientific and technical advisory committees and formal peer reviews. Often, when missions are selected, neither OSSA nor scientists have invested sufficient resources to reliably compare cost and scientific productivity with less costly mission alternatives. (Experienced space scientists and program managers have documented that it is almost impossible to make reliable estimates of the cost of a scientific spacecraft until the first 10% of the development cost has been spent.)[9] Lacking such information, scientists must rank the proposed missions according to the projected scientific performance without being able to gauge the impact of the mission's budget on the scientific program as a whole.

Since projected performance, not budget, is the overriding criterion for selection, scientists have had little incentive to build their instruments to minimize costs or to make conservative budget and performance estimates. Instead, they typically design to the maximum performance feasible within the spacecraft envelope.

The trend toward bigger missions is built into the national political process as well. The Office of Management and Budget (OMB) in the Executive Branch has traditionally held OSSA to a fixed number of new starts per year (usually one), rather than a fixed budget envelope. With such a constraint, it is of course to OSSA's advantage and the perceived advantage of the scientific community to make each mission as big as possible. Moreover, Congressional staffers and several former NASA and OMB officials have pointed out[10] that it is usually no more difficult (and perhaps easier) to get a new start for a mission costing $1 billion than for one costing $300 million. With a billion-dollar mission, the political pressures that can be brought to bear by a broad-based constituency of scientists, universities, NASA centers, and the aerospace industry may be great enough to break through the same political obstacles that can impede a $300 M mission.

Given such a constraint on the number as opposed to cost envelope of new starts, space scientists working in a given discipline are motivated to press for the biggest possible mission, since they perceive that they will have only one such opportunity in a decade or more. This perception creates what may be called the "Christmas-tree effect" in which scientists working in a given subdiscipline ornament the mission with a substantial array of scientific instrumentation because they believe that the mission will provide their only career opportunity to fly their instruments. The Christmas-tree effect then becomes a self-fulfilling prophecy, because it drives the cost of the mission so high that OSSA will not have the resources to build another mission for that subdiscipline for a decade or more.

The perception of an opportunity for a big mission can blank out the possibility to have smaller missions in the same field. The existence of smaller missions would remove the political "cruciality" argument that is perceived to be vital for starting a large or even moderate mission of a given type: *i.e.*, such a mission is "crucial" for the survival of an important discipline within space science.

Sometimes this strategy works, but sometimes it backfires. For example, in the early 1980s, US infrared astronomers turned down an opportunity to build an instrument for the powerful Infrared Space Observatory (ISO) that will be launched in 1994 by the European Space Agency. They did so because they believed that NASA would propose a new start on the Space Infrared Telescope Facility (SIRTF) in 1984 or 1985. According to the OSSA Strategic Plan, NASA will propose SIRTF as a new start in 1993 or 1994, and in fact SIRTF will not be launched before 2000. Similar circumstances in part prevented the US from flying a small mission to Halley's comet, leaving the principal scientific returns to European, Japanese, and Soviet investigators. As these examples illustrate, the concentration of resources in a single large mission can subject a scientific subdiscipline to a "programmatic single-point failure mode" that can be as devastating as a launch failure.

The concentration of scientific resources in a single mission reaches a new level with the planned large Earth Observing System (EOS) platform, which is particularly vulnerable to the dangers described above.

We believe that one of the most powerful forces driving OSSA to concentrate its resources in missions of increasing size and complexity is a lack of confidence, by both scientists and NASA managers, that Congress and the Executive Branch would be willing to support a program of space science that is not tied primarily to individual large missions.

The Culture of Risk-Avoidance

Failure is intrinsic to experimental science; indeed, it is an essential mechanism by which scientists and engineers learn and make progress. Of course, NASA cannot eliminate risk from its space science programs. Yet, one of the most striking recurrent comments that we heard from space scientists was that "NASA doesn't know how to take risks anymore." What does this statement mean?

Beginning in the 1970s, NASA made a fundamental shift in its approach to risk management. In the 1960s and early 1970s, NASA often took technical risks (*e.g.*, by launching complex missions to land on Mars) but was careful to avoid programmatic risk (*e.g.*, by flying two or more spacecraft to accomplish a single objective). See the following chapter by Stern and Habegger for a fuller discussion of this shift.

In the 1960s and early 1970s, when NASA flew the various multi-spacecraft space science programs, it managed these programs according to a "failure-resilient" strategy. OSSA typically supported the construction of clones of science payloads so that there would be a backup in case a mission failed. That strategy is not as expensive as it might seem, because most of the cost of building a scientific payload can be attributed to engineering and setups for fabrication and testing. But in the 1970s, under pressure to maintain a broadening scientific base in a budget-constrained environment, NASA changed to a "success-oriented" strategy, in which technical risk was to be reduced by increasing quality control of a single payload rather than insured against by building a backup. It is not at all clear, however, that the success-oriented strategy provides a gain in productivity, because the incremental cost incurred from trying to avoid technical risk can easily exceed the incremental cost of building a clone. *This replacement of a failure-resilient strategy by a success-oriented strategy is tantamount to replacing technical risk by programmatic risk.*

This shift in risk-taking has been exacerbated by the inability of both NASA and the scientific community to sacrifice breadth across all subdisciplines for depth in a few. OSSA now supports many new subdisciplines (*e.g.*, space oceanography, X-ray astronomy, materials sciences) that did not exist 25 years ago. In the 1960s and 1970s, NASA supported programs to fly a series of missions in a given core discipline. Today, NASA usually devotes a single mission to each subdiscipline. Accordingly, eight Surveyor lunar missions have been supplanted by one large Galileo mission; four Orbiting Astronomical Observatories have been supplanted by one large HST; eight Orbiting Solar Observatories have been supplanted by one planned Orbiting Solar Laboratory; and so forth.

This shift in strategy is a key obstacle to cost containment of scientific programs; moreover, it ultimately results in a space science program that is more vulnerable to failure.

One important element of this technical "risk management" is the ever-growing demands for documentation, which have become a substantial fraction of the mission cost. For example, if a scientist wants to use a component or subsystem that does not conform to a NASA standard, he or she must obtain a waiver; but the difficulty and cost involved are likely to be so great that the scientist may elect not to fight the system, even if meeting the
standard impairs the scientific performance or increases the overall cost of the mission.

This problem stems largely from the fact that the people responsible for setting safety and reliability standards have no stake in optimizing the scientific productivity or cost of the mission -- only in avoiding risk.

Moreover, no amount of documentation, study, and reviews can completely remove risk. As the HST and Galileo experiences painfully illustrate, this approach to risk management is a poor substitute for actual experience and independent testing.

Simple arithmetic tells us that if space scientists could achieve, say, a 50% cost savings by deliberately accepting a 20% risk of failure per mission, net scientific productivity would rise by 33%. (For example, instead of spending $200 M for one mission and its $40 M launch, with an expectation of nearly 100% probability of success, spend $240 M for two missions and two launches, each with 80% success probability, gives an expectation of 1.6 successful missions.)[11]

In fact, a strategy of cutting costs by deliberately accepting technical risk is likely to *decrease* the risk of total failure. For example in the above example there would be an expectation of 0.64 for two successful missions and 0.96 for at least one success.

The deliberate acceptance of technical risk will be unacceptable for scientists, however, if the risk applies to a single major mission upon which one's career and even the health of an entire discipline depends.

There is ample evidence that the culture of risk-avoidance, the "single new start" dilemma, the Christmas-Tree effect, and the concentration of resources in large missions are intimately coupled and mutually reinforce each other. Breaking this cycle is thus an essential part of the cure for the problems of the Nation's space science program.

Non-resilient Budget Planning

For the OSSA Strategic Plan to work, two assumptions must prove valid: (1) the assumed 10% real annual growth of the OSSA budget must materialize; and (2) the missions must be carried out at the estimated costs.

How valid are these assumptions, and what will happen if either one fails? The total NASA budget request for 1991 was approximately $15.1 billion, an increase of about 23% over the 1990 budget. Congress actually appropriated $13.9 billion, providing an increase of 14%. The OSSA request budget was protected, however, and actually increased by 19%. This increase was largely targeted toward the new Earth Observing System (EOS) initiative. If the OSSA budget does not continue to grow, OSSA cannot carry out much of its Strategic Plan. An equally serious worry is whether OSSA can hold the costs of its missions to the current projections. That has not been the case in the past decade.[12] For example, the development cost (in constant dollars, excluding launch) of the Hubble Space Telescope increased from approximately $0.5 billion when it was started in 1977 to more than $2 billion today.

A standard strategic planning device is to fit the cost curves of various planned activities within a projected funding envelope, as illustrated in Figure 1. From such a diagram, it is easy to see what will happen if either (1) the envelope does not grow as assumed; or (2) one of the major missions exceeds its budget by a substantial factor. Because expenditures cannot exceed the funds appropriated in any given year, the result is the "squirt effect": future missions are pushed downstream by years. In fact, the squirt effect has been just as responsible as the Challenger accident for the stretch-outs of the major missions that were planned by OSSA for the 1980s.

The squirt effect is driven by a nonlinear, positive feedback amplification because the runout funding of a mission (the area between two curves) increases substantially if the mission is stretched out. Mission development normally requires a fixed minimum rate of spending to support a base of technical expertise, whether or not the people comprising the base have the financial resources to carry out their assigned tasks. For example, if sufficient funds are available, a contractor will purchase some parts from a specialized subcontractor for efficiency, but with fewer funds the contractor may manufacture the parts in-house thus delaying progress. Thus, a stretch-out in one mission directly impacts the productivity of the money spent for that mission. Moreover, fitting the increased net funding within a fixed costenvelope mandates stretch-outs in other missions, propagating the effect.

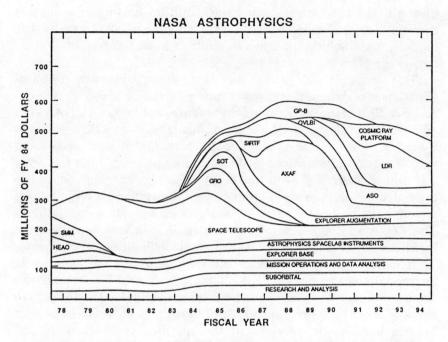

Figure 10.1: A Typical Planning Diagram for NASA Space Science Projects. Source: NASA, *The Crisis in Earth and Space Science*, (1986).

The productivity-cost associated with schedule stretch-outs would seem to argue that it would be better to delay the start of a mission than to risk a stretch-out. However, the need to prove critical technology is a strong argument for early funding of a mission. When concerns regarding critical technology arise during advanced stages in the development of a mission, delays become very costly. The result is that commitments are made to start missions with funding curves that are "back-end loaded." With a number of such "slow starts" – narrow wedges competing within an uncertain funding envelope – the strategic plan is most susceptible to the squirt effect.

The federal budget process contributes to the dynamic of the squirt effect. The development of space science missions requires some 4 to 5 years, even under optimal circumstances. Yet OSSA cannot count on its annual budget more than a year or two in advance. Thus, it is impossible for OSSA to predict the impact on future missions resulting from a decision to augment or stretch out the funding of a mission currently under development. In the absence of stable multi-year funding, there is little incentive to make rational trade-offs based on long-range planning.

This dynamic is exacerbated by tension between political imperatives and the imperatives of rational strategic planning. On the cost side, it is standard Washington practice, not limited to NASA, to underestimate the cost of programs being sold to OMB and the Congress. On the budget side, NASA feels that it must present an annual request which is optimistic regarding program costs (i.e. low) and future budget growth (i.e.high). That is, programs must be presented as inexpensive in order to be approved and a large budget growth must be requested as a starting point in budget negotiations. The OSSA strategic plan is constructed in this way, and thus the inefficiencies of the squirt effect are implicitly built into the process.

Recently, Congress has instituted a system of multi-year appropriations to OSSA's CRAF/Cassini and AXAF programs. To track or manage these missions Congress has legislated fixed-objective checkpoints in each program, at which time the next two-to-three year funding block is be approved if the specified cost, schedule, and performance targets have been met. We recommend that Congress institute such a system of appropriations as the norm for major OSSA missions.

Inappropriate Linkage to the Manned Space Flight Program

The exploration of space by human beings is an historically important endeavor. However, one unfortunate consequence of the post-Apollo decline in NASA's programs was an effective near-prohibition on the use of Americans in space for exploratory purposes.

In turn, when NASA fought to hold onto a manned spaceflight capability, its only available avenue was to couple manned spaceflight to space transportation.[13] In order to ensure the very high launch rate required for the Shuttle to achieve its intended economic advantage, NASA decided in the 1970s to phase out use of expendable launch vehicles.[14] Thus, other space endeavors were effectively held hostage to the Shuttle's success in order to preserve the future of the manned space flight program.

Surely, in some instances, space science missions require close coupling to manned spaceflight. An undisputable example is human biomedical investigations. Arguable examples include spacecraft servicing and repair, field geology on planets, and materials science experiments.

Still, however, there are certain drawbacks to the coupling of space science to the Shuttle.[15] Unlike expendable rockets, the Shuttle is constrained to low Earth orbits, penalizing the productivity of many scientific missions. For example, because of the viewing constraints, the restriction to Shuttle-compatible orbits reduces the observing efficiency of the HST to roughly 30% (compared to roughly 90% for the International Ultraviolet Explorer (IUE) observatory in geosynchronous orbit). Moreover, satellites in low orbit may require the TDRSS communications satellite system for ground communications, introducing serious logistical problems (*e.g.*, competition with military requirements), new failure modes, and greatly increased operational complexity and cost compared to operation in geosynchronous orbit. Further, Shuttle launches are much more vulnerable to schedule delays than ELV launches in part because the safety constraints are more severe, and delays in one launch affect all subsequent programs owing to its being a reusable vehicle.

The launch of scientific payloads by manned spacecraft also introduces an extra level of safety requirements, design restrictions, and documentation that substantially increase the cost of the payloads and may constrain scientific performance. The added cost of safety requirements for scientific operations associated with manned operations was evident to scientists working on Spacelab payloads. For example, one university-based scientist reported to us that he lost an opportunity to re-fly an experiment already built and successfully flown on a previous Spacelab flight, because it would cost OSSA more than $2 M to meet new manned spaceflight safety requirements introduced after the Challenger accident. The original cost of building his instrument was $1 M. Among the new requirements was the demand that his working, flight-proven, and calibrated instrument be dismantled and inspected, part-by-part.

Another example of the inefficiency of launching science payloads with Shuttle is provided by the successful COBE mission. This experiment was originally built for a Shuttle launch but was re-configured for launch by a Delta rocket after the Challenger accident. In the process, the satellite weight was reduced from 10,500 to 4,860 pounds *with no sacrifice of scientific performance.* (Indeed, COBE has delivered spectacular results.) The experience with the

COBE satellite clearly shows how the requirements of compatibility with the manned spaceflight program can add weight, cost, and complexity to the payload.

To its credit, OSSA has re-introduced the purchase of ELVs to launch some scientific satellites. However, it is premature to conclude that inappropriate linkage of scientific payloads to the manned missions has been completely removed. Decisions to rely on STS are still being made today. For example, OSSA still plans to re-use the spacecraft that will carry the Extreme Ultraviolet Explorer (EUVE) experiment by retrieving this satellite with Shuttle and replacing the EUVE telescope with other experiments, first the X-ray Timing Explorer (XTE) and then the Far Ultraviolet Spectrometer Explorer (FUSE). The decision to re-use the EUVE spacecraft for other Explorers is dictated by a false economy: the NASA astrophysics division doesn't have to buy two more spacecraft if the original spacecraft bus is reused, and the shuttle retrieval launches are "free" to OSSA. Of course, the Agency may wish to develop on-orbit payload replacement technology and experience; however, there is no *scientific* imperative for the strategy. There are, however, scientific penalties. The efficiencies of the experiments will be reduced by a factor of two or more because of the low earth orbit required for Shuttle tending. Further, successful, functioning payloads will be forced off the spacecraft by succeeding instruments even though they are likely to be capable of delivering valuable science for several more years.

Paradoxically, if the launch costs associated with refurbishing these satellites from Shuttle were added, it would be much cheaper for NASA to build three separate spacecraft and launch them independently from Delta rockets. (The actual cost of a Shuttle flight is approximately $400 M, compared to $50 M for a Delta launch). However, OSSA has no incentive to make a rational choice because of NASA's budgeting procedure, in which the OSSA budget is not charged for launch and operations costs. Thus, for OSSA the launch and operations costs associated with manned refurbishment are "free," whereas the costs of building duplicate spacecraft are real. No cost-performance analysis is valid if launch and operations costs are excluded. Until the NASA budgeting process is structured so as to make OSSA responsible for launching and tending scientific experiments, the SESAC Report's goal of selecting launch vehicles primarily on grounds of scientific return will remain elusive.

The Erosion of Scientific and Technical Expertise

A serious long-range threat to space science, indeed to the future of NASA itself, is the failure to train a sufficient number of space scientists, instrumentalists, and engineers. The Agency recognizes that it faces a scientific manpower shortage and is taking action to rebuild the base, but the current strategy does not go to the root of the problem.

Of course, NASA's difficulty in finding enough highly trained engineers and scientists is symptomatic of a nationwide problem in training scientific and technological manpower.[16] NASA is not solely responsible for the problem, nor can the Agency solve it alone. NASA does, however, have a unique role to play in contributing to the solution, particularly as regards its own manpower needs.[17]

In the 1960s, there was a natural path for the training of a space scientist. As a graduate student, a scientist would learn to build space instrumentation, fly it on a rocket, and analyze the data for his or her Ph.D. thesis. Then, as a junior faculty member, the scientist would gain experience in the management of more complex systems by working with a team of scientists to build and operate a small satellite mission. The most capable of such scientists would go on to become leaders of the teams that built instruments on the major missions awaiting launch in the next few years. Other entry paths contributed significantly, such as the transfer of people from experimental high energy physics into space science.

In an important article, "A Changing University Role in Space Research," J. D. Rosendhal and L. J. Lanzerotti[18] describe how the shift from small to large scientific payloads is leading to fundamental changes in the ways that universities are training scientists and participating in space science research. They point out that with the increasing size, complexity, and 10 to 15 year lead times of the major missions that now dominate space science, it is no longer feasible for a young scientist to participate in all aspects of a space experiment within the normal duration of graduate study or a junior faculty appointment. As a result, they suggest, the role of universities in space science will shift heavily toward the analysis of data from major missions. With the loss of the projected opportunities for university researchers to participate in the design and construction of a wide spectrum of space experiments associated with the Shuttle, the remaining opportunities will lie largely in suborbital (rocket, balloon, and aircraft) programs. The major missions will be developed and operated by a small number of large research teams involving university-industry-government consortia.

The Rosendhal-Lanzerotti article is an insightful and accurate description of the programmatic direction of space science research today. Taken as a warning, the article is a valuable contribution; but taken as a policy recommendation, it would be a recipe for disaster. Today and for the foreseeable future, universities are the only institutions capable of producing new space scientists. Certainly, interpretation and analysis of data is an essential aspect of space science, but people cannot be trained to invent and build new space experiments by sitting in front of computer terminals.

The missing link in their scenario is the mechanism for training people to build and manage the development of space science projects. Lanzerotti and Rosendhal recognize this problem and suggest that it may be solved by importing scientists from other disciplines, such as experimental physics, which have contributed to space science all along. However, our conversations with high energy physicists revealed the flaw in this logic. One laboratory director answered: "We can't find enough technically qualified people to work on our own projects. We were sort of hoping to recruit some people from space science."

NASA cannot count on other disciplines to solve its manpower problems. Indeed, the Agency should be an exporter, not an importer, of such people. Major missions are not an effective mechanism for training new scientists and technicians. As the size of a mission increases, it becomes too big for most universities to handle. An increasing fraction of the funds expended must go to aerospace industry, which has unique capability to build and manage the large and complex subsystems involved. There are heavy penalties, however. First, a decreasing fraction of the money spent for space science goes to the training of new scientists; and second, space science loses the flow of innovation that traditionally has been provided by university research groups. In the long run, the aerospace industry suffers too, because it loses the flow of new talent that the universities alone can provide.

Without a dedicated, near-term effort to decrease the timescale of a typical space science mission cycle (from inception to data return) and a commensurate increase in the number of missions, the Nation faces the prospect that most of its experienced space scientists will retire without having trained a fully-versed replacement generation. Such a scenario, which in fact lies less than 10 years ahead,[19] may result in an *increased* rate of space science mission failures, since the less-experienced replacement generation will have to get its training in real time.

Toward a More Resilient Space Science Program

We now make several recommendations following from the discussion above. The implementation of such recommendations should lead to a more robust space science program.

A More Stable Mix of Scientific Missions

OSSA presently plans one major mission each year, one to three Spacelabs each year, one or two Small Explorers (SMEX) each year, and a moderate mission once every two or three years. Estimating (for sake of illustration) the cost of SMEX missions at $30 M each, moderate missions at $120 M each, and major missions at $600 M each, we estimate that OSSA must spend some $750 M per year for free-flying satellites (excluding launch and operations costs). This sum will be divided approximately as follows: SMEX - - 12%; moderate missions -- 6%; major missions -- 82%. Even if a new major mission were started only once every two years, the major missions would still consume 70% of the total OSSA budget for free-flying satellites. This unbalanced mix is the main reason for instability and brittleness in the OSSA space science program. We suggest that NASA, OMB, and the Congress instead work toward a distribution of expenditures for scientific satellites that is weighted more heavily toward moderate and smaller missions. For example, with the same annual budget for new missions, OSSA could fly eight SMEX flights and two moderate missions each year, and one major mission every three years.

With such a distribution, a vigorous program of space science is possible. The strategy of shifting resources toward missions of smaller size and shorter lead times permits more extensive involvement of university groups and students in all aspects of the program. This strategy will also be more stable and immune to the "squirt" and "Christmas-tree" effects than the present one, would foster a robust set of training opportunities for young researchers, and would provide a scientific program more resilient to occasional failures, which are unavoidable in experimental science.

It will be a challenge for NASA to return to the high launch rate implied by the shift to a greater number of small and moderate missions. For example, this goal may require that NASA upgrade its facilities at the Kennedy Space Center or use foreign launch facilities. It will, however, be an even greater challenge for NASA to regain the capability to launch and operate these payloads in a routine way.

Learning from Japan

The Institute of Space and Astronautical Sciences (ISAS) of the Ministry of Education of Japan has launched a series of missions to observe cosmic X-ray sources. The Japanese have already flown three of these satellites on roughly 4-year intervals, and are now in the process of preparing the fourth, ASTRO-D, for launch in 1993. As a result of the coincidence of the success of the Japanese experiments and the hiatus of launches of US X-ray experiments during the past decade, the Japanese group has taken the lead in X-ray astronomy, a field pioneered by NASA.

To be fair, the capabilities of the ASTRO missions have been modest compared to the most recent NASA mission for X-ray astronomy, the very successful *Einstein* satellite that was launched in 1979. Indeed, the first three ASTRO instruments have been similar in design and comparable in capability to instruments flown by NASA in the mid-1970s. In contrast, the instruments on the ASTRO-D mission incorporate major advances in design, and ASTRO-D will likely be one of the most powerful X-ray telescopes flown until NASA launches the AXAF mission in 1998-99. Even so, the major innovations of the ASTRO-D experiment, *e.g.*, the telescope mirrors and the X-ray CCD's, were originally developed by US scientists and will be implemented as a collaboration between ISAS and NASA.

ISAS has been able to take the lead in X-ray astronomy, not by their superior scientific skills, but by their consistent ability to build a scientific satellite at a cost estimated to be 1/3 to 1/4 of the cost that the U.S. requires to build a satellite of comparable performance. For example, the ASTRO-D experiment will cost the Japanese government approximately $50 M, whereas a NASA mission of comparable performance would be budgeted approximately $150 -- $200 M. How does ISAS do it?

To understand this, we interviewed Dr. Yasuo Tanaka, the leader of the Astrophysics group at ISAS. What he told us suggests that the Japanese space scientists operate under a different set of rules than US scientists building missions for NASA. The fundamental difference is that the Japanese scientists have total responsibility for the management of the project, and also total accountability. When a mission is approved, the Ministry of Education funds ISAS according to two boundary conditions: fixed total cost and fixed time schedule (approximately five years). According to Dr. Tanaka, once a commitment is made, these conditions are practically immutable.

We asked Dr. Tanaka what would happen if the project went over budget or was not ready by the scheduled launch date, and he replied that his group would not allow either option, as they would have a severe impact on the entire ISAS program. If they foresee a cost overrun, they would compromise the specifications. In this respect, Dr.Tanaka believes that scientists are smart enough to find a way out without a major compromise of the scientific goals. If the group was behind schedule, they would work overtime to catch up.

Who takes the management responsibility? Dr. Tanaka replied, "In the case of an X-ray astronomy mission, a senior X-ray astronomer at ISAS would be assigned as the project manager." Budget assessment? "The project manager is required to do it, *i.e.*, I do personally for my mission." Risk assessment? "ISAS has accumulated a great deal of expertise and we are heavily assisted by the ISAS engineers who are very good at risk
assessment. Based on the ample evidence that they provide, I and my group make our own decisions on optimal choices between risk, cost, and scientific performance." Yet, Dr. Tanaka emphasized, "While this management style is efficient and cost-effective, it works because the scale of ISAS missions is modest and the staff is capable of giving the right technical judgement."

Finally, we asked Dr. Tanaka about how he deals with subcontractors in the Japanese aerospace industry. He replied, "We begin with guidelines rather than detailed specifications." Why? "Industry always prefers to set contract on detailed specifications. However, our contractors are well-aware that specifications evolve as the work proceeds, and they cooperate flexibly with us in this style."

The contrasts between the Japanese and NASA styles of building space experiments are profound and certainly deserve deeper study. Would the Japanese style work in the US? Possibly. There are obstacles, most serious of which may be that very few US space scientists have had the opportunity to gain the kind of project management experience that the Japanese scientists have. But such a management style is not completely foreign to US science and technology. US physicists have built large high energy particle accelerator facilities and have largely retained responsibility for management of the projects. Further, some US aerospace firms have used such techniques in their "skunk works" and are now broadly adopting "total quality management" techniques based in part on Japanese strategies.

There are other serious obstacles as well, some of which are beyond NASA's ability to control. The two most intractable problems may be: (1) government regulations that impair the ability of project managers to revise procurement strategies as conditions warrant; and (2) the annual congressional appropriations practice that impairs NASA's ability to make realistic long-range commitments. The Japanese and European space science agencies do not suffer these problems.

Reforming the Mission Development and Management Process

NASA's space science program should provide quicker access to space with small and moderate missions. In contrast to the larger missions, many smaller missions could be fit into the present budgetary envelope. However, this can occur only if OSSA finds more effective ways to contain the cost and schedule of each mission.

There are four crucial elements to a successful cost-containment strategy: (1) the scientific community must take into account mission costs in establishing its scientific priorities; (2) the costs must be realistic; (3) full authority and accountability for completing the mission within cost and schedule must be vested in a tight management team including the project manager and the principal scientist; and (4) funding must be provided as required to support the schedule.

OSSA has already begun to introduce incentives for cost containment into the Explorer Program selection process by supporting a greater number of missions for the definition phase (Phase A) and by conducting a second competition to select missions for completion. Such a strategy is an improvement over past practices. To ensure that this strategy is successful, OSSA should also: (1) include mission costs and their impact on future missions as criteria of the peer review process in both the Phase A and Phase B competitions; (2) make it clear that the management team will be held to its budget, even if it becomes necessary to de-scope the scientific performance of the mission.

If scientists knew from the outset that cost control would be a major factor in selection, they would devote more energy to developing creative strategies for building low-cost missions. If the management team were accountable for the scientific return, and knew that it would be held to the proposed budget, even at the possible cost of scientific performance, the incentive to propose unrealistically optimistic budgets would be removed. Another vital requirement for cost control is that missions selected for development (Phases C and D) must have their critical technologies under control. Schedule delays caused by unforeseen technical problems are very costly once the mission has entered these phases. In order to meet this requirement, NASA must be prepared to invest more money for technology development and design (in Phases A and B).

Project managers should optimize a mission by balancing scientific performance, cost, schedule, and risk. In order for this process to work effectively, the decision-making responsibility must be vested in a management team, led by a single individual of demonstrated competence, that will suffer the penalties as well as realize the benefits from such decisions. The management team must have direct authority over all personnel and resources that are vital for completion of the project, and must be provided multi-year appropriations to efficiently manage the master schedule.

At present, authority for project management is too diffuse. For example, when full authority for quality control is removed from the management team, either by reporting relationships or by paper requirements, that authority will be exercised by people who have little knowledge of the context of the specific mission and no stake in minimizing costs or maximizing scientific performance. If the project manager lacks the necessary financial resources and authority, he cannot be held accountable for the costs and delays that can be expected when such authority is imposed from outside the project.

In mission development, priorities for scientific instruments and spacecraft capabilities should be established to facilitate de-scoping to control costs in the event of technical difficulties. To its credit, the OSSA Planetary Division has employed such a strategy with the painful but precedent-setting de-scoping of the original Venus Radar and Mars Observer missions.

There must, of course, be mechanisms to detect and correct failures in project management. But, we have heard ample testimony that such mechanisms are operating *even in the absence of any evidence of failure*. Indeed, mechanisms introduced to avoid rather than correct failure -- oversight teams, project reviews, documentation requirements -- have become so intrusive that they preclude performance optimization and effective cost and schedule control. The appropriate way for NASA to introduce oversight into mission management is to establish agreements with the management team for milestones and review procedures at the beginning of the development phase. No additional oversight or documentation procedures should be implemented if these milestones are met.

Conclusions

Although the health of the NASA space science program has improved in recent years, serious problems persist. We have examined the root causes underlying these problems and find them to be both endemic and mutually reinforcing.

Among the problems, we have found that (1) strong institutional and political forces favor big missions over small ones; (2) unrealistic budget planning removes flexibility, resulting in reduced productivity and increased costs; (3) a culture of technical risk-avoidance and programmatic risk-taking exists which is detrimental to an innovative, resilient, and productive space science program; (4) pressures persist which encourage inappropriate linkages of scientific experiments to the manned program; and (5) a serious shortage of people with experience in building scientific spacecraft and managing their development exists now, and is growing more severe with time.

These problems were not generated by NASA alone. Congress, the Executive Branch, and the scientific community all share the responsibility for the present situation. All must cooperate to improve it.

We suggest that key elements in a strategy toward achieving a more durable and productive space science program are: (1) a shift in the distribution of space science resources favoring a greater frequency of small and moderate missions; (2) increased authority and responsibility for scientific program managers, particularly with regard to the overall structuring of mission launch and flight operations plans; and (3) multi-year appropriations for space science projects.

Notes

1. The authors thank L. Acton, J. Alexander, J. Arnold, P. Banks, J. Brandt, R. Brunner, G. Briggs, R. Byerly, C. Canizares, T. Donahue, L. Esposito, R. Giacconi, D. Helfand, S. Holt, R. Konkel, L. Lanzerotti, J. Linsky, R. Mushotzky, R. Roney, B.

Savage, M. Shull, H. Smith, Y. Tanaka, A. Wheelon, and several other colleagues for their valuable comments on a draft of this paper. The opinions expressed here are entirely the responsibility of the authors.

2. *Report of the Advisory Committee On the Future of the U.S. Space Program*, Washington, D.C.: U.S. Government Printing Office, 1990.

3. *Office of Space Science and Applications Strategic Plan*, NASA, 1990.

4. *The Crisis in Space and Earth Science*, Space and Earth Science Advisory Committee of the NASA Advisory Council, 1986.

5. Lanzerotti, L.J., and J.D. Rosendhal, *Physics Today*, May 1988; Giacconi, R., in *Space Policy Reconsidered*, R.Byerly, Jr., Ed., Boulder, CO, Westview Press, 1989; Brown, R.A., and R. Giacconi, *Science*, 238, 617 1987; Morgan, W.G., *Issues in Science and Technology*, 72, Spring 1989.

6. Fisk, L.A. personal communication, 1989.

7. Augustine, N.R. *Augustine's Laws and Major Systems Development Programs*, American Institute of Aeronautics and Astronautics, New York, 1983.

8. Bless, R. *Issues in Science and Technology*, 67, Winter 1988-89.

9. Hearth, D.P. *NASA Program management and Procurement Procedures and Practices*. Testimony before the Subcommittee on Space Science and Applications, Committee on Science and Technology, U.S. House of Representatives. 24 June 1981, U.S. Government Printing Office, 82-309, 1981.

10. Moore, D.H., in *Space Policy Reconsidered*, R. Byerly, Jr., Ed. Boulder, CO, Westview Press, 1989.

11. This illustration assumes independent failure modes: If two identical spacecraft were built they might have similar failure modes which means that the calculations would be much more complicated.

12. *Space Exploration/Cost, Schedule, and Performance of NASA's Galileo Mission to Jupiter*, U.S. General Accounting Office, GAO/NSIAD-88-138FS, Washington, D.C., 1988; *Space Exploration/NASA's Deep Space Missions are Experiencing Long Delays*, U.S. General Accounting Office, GAO/NSIAD-88-128BR, Washington, D.C., 1988; Moore, D.H., in *Space Policy Reconsidered*, R. Byerly, Jr., Ed. Boulder, CO, Westview Press, 1989; Hearth, D.P. *NASA Program management and Procurement Procedures and Practices*. Testimony before the Subcommittee on Space Science and Applications, Committee on Science and Technology, U.S. House of Representatives. 24 June 1981, U.S. Government Printing Office, 82-309, 1981.

13. McDougall, W. *The Heavens and The Earth: A Political History of the Space Age*. Basic Books, Inc., New York, 1985.

14. Murray, B. *Journey Into Space: The First Three Decades of Space Exploration* New York. Norton Press, 1989. See also Pielke, et al, Chapter 14, this volume; Logsdon, J.M. *Science*, 232, 1099, 1986; Wheelon, A.D. *Space Policy: How Technology, Economics, and Public Policy Intersect*. Working Paper No. 5, Program in Science, Technology and Society. Boston, Massachusetts Institute of Technology, 1989.

15. See endnote 7.

16. *The Civil Service Workforce: A Report to Management, Fiscal Year 1988*, NASA Headquarters, Personnel and General Management Office, Washington, D.C., 1988, p.65; Windall, S.E., *Science*, 241, 1740, 1988; Atkinson, R.C., *Science*, 248, 425, 1990.

17. *Toward a New Era in Space: Realigning Policies to New Realities,* Committee on Space Policy, National Academy of Sciences and National Academy of Engineering, Washington, D.C., National Academy Press, 1988.

18. Rosendhal J.D. and L.J. Lanzerotti, *Issues in Science and Technology,* V, 61-66, 1989.

19. Stern, S.A., Konkel, R., and Byerly, R. "Demographics of NASA-Funded Space Research Principal Investigators", *Space Policy,* 6, 1990.

Chapter 11
AND THEN THERE WAS ONE:
THE CHANGING CHARACTER OF NASA'S SPACE SCIENCE FLIGHT PROGRAM

S. Alan Stern and M. Jay Habegger

E PLURIBUS UNUM

Introduction

In the late 1960s and early 1970s, NASA launched a series of four spacecraft that shared the name Orbiting Astronomical Observatory (OAO). Although some of the scientific instruments aboard each spacecraft were unique, there was a substantial degree of commonality among the spacecraft support systems, instruments, and mission objectives. Further, instruments on one spacecraft were largely designed to complement observations made by others. Two of the OAO spacecraft failed, but the other two successfully returned ground breaking astrophysical data. The OAO program has long been considered a major success.

In contrast to the series of OAOs, consider the collection of "Great Observatories" that are being built and launched today. The *Hubble Space Telescope* (HST), the first of these observatories, was launched aboard the space shuttle Atlantis in April, 1990. The *Gamma Ray Observatory* (GRO), launched in April, 1991, is the second Great Observatory.[1] Designed to make both imaging and spectroscopic observations, HST supports the same disciplines within the astronomical community that the earlier OAOs supported. There is only one *Hubble*, however. Indeed, each of the Great Observatories is a unique spacecraft supporting a substantially different community of investigators. Furthermore, each of the Great Observatories is a separate program composed of a *single* flight mission.

Anecdotal evidence about programs such as OAO described above, is often used to support the argument that a *qualitative* difference exists between NASA's approach to space science missions and the approach to missions and flight programs in NASA's "golden age".[2] This anecdotal evidence is sometimes used to claim that the present space science program is more risky than it would be if each program contained multiple missions. In this Chapter, we *quantitatively* examine this claim to assess its viability, and examine the impacts of the *actual* trends in space science missions that have taken place over the past three decades.

The Record

In order to draw quantitative conclusions about how the NASA space science program has changed, it is necessary first to develop a complete database of space science missions that have been supported by NASA's OSSA and its predecessor. The individual missions in the database can be analyzed, grouped into programs and arranged chronologically.

One issue that confronted us in the preparation of this database was which missions to include. We wished to analyze only the character of the *unmanned* space science program, and not the pattern of the entire space science and applications effort. Missions that were manned or human-tended (e.g. Skylab, Spacelab) were excluded, as were missions that were primarily directed at proving a technology or demonstrating engineering capability, such as early experiments in satellite communications (e.g. the Echo series). Missions designed to provide applications services were also excluded, such as the series of Tiros weather satellites and the Advanced Communications Technology Satellite currently under construction. Also excluded from out study were balloons and sounding rockets. Programs sponsored by a foreign country with limited NASA involvement, such as the largely-German Roentgen Satellite (ROSAT) and the cooperative U.S.-German HELIOS program, were omitted from our study because the programmatic decision-making process in designing these programs was strongly influenced by foreign participation.

A second issue that arose in the preparation of our database derived from the fact that we had to develop a broadly applicable set of criteria for determining if a set of missions constituted a single program, rather then a collection of independent missions. The operational standard we adopted was that a significant degree of commonality must exist between individual missions in terms of scientific objectives, spacecraft hardware, and organizational infrastructure in order for the sequence to be classified as a single program. We investigated each mission individually to determine the proper classification, rather than just relying upon mission titles. All of the spacecraft in a program composed of multiple missions need not be identical. Our intent in this classification is to study both discipline-specific and pan-discipline trends in the overall space science program.

Two simple examples may clarify the process of classifying missions into programs. First, consider the Surveyor series of lunar lander spacecraft that were launched in the late-1960s. All of the Surveyor spacecraft were similar, and the mission of each was to provide data on the Moon's surface. Classification of the Surveyor missions as a program of seven spacecraft supporting planetary scientists is not difficult. Even though the later missions in the program carried more sophisticated instruments, the series of seven spacecraft had common goals and management and a significant amount of hardware similarity.

Unfortunately, not all missions can be so easily classified. As a second example, consider the series of Explorer missions that stretches from the early-1960s through the present. Despite the common name of the missions, the series of Explorer missions cannot be considered a single program containing a large number of missions. The breadth of the missions in the series of Explorer missions ranges from earth science to space physics. Separating the Explorer missions into programs was accomplished by obtaining information on each mission and determining the degree of commonality between missions. Some of the Explorer missions were unique spacecraft with unique goals, while others were part of a series of Explorer missions.

We now describe what our findings reveal about the space science program. Figure 11.1 illustrates the portion of the OSSA science missions carried out in each of the five conventional research disciplines of the civil space science program. Figure 11.1 reveals that with the exception of life sciences, each discipline area has received approximately the same percentage of flight missions. The category labeled "other" is primarily composed of OSSA missions designed to study micrometeorites. Of course, the raw flight rates in Figure 11.1

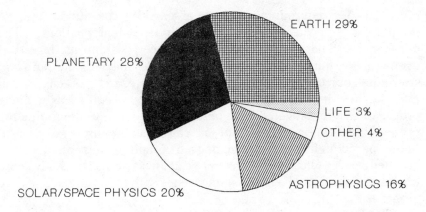

Figure 11.1: Unmanned NASA Science Missions, 1960-2000. The portion of the total number of OSSA science missions that have been carried out in each research discipline. Sources: NASA. *TRW Space Log*, V. 23, (1988).

do not reveal the distribution of costs and data return by each discipline, but this was beyond the scope and intent of our study.

In Figure 11.2a we show the absolute number of U.S. civil space science missions launched by year from 1960 to 1990 and the relative proportion of these missions that were part of a program containing multiple missions. This figure also includes OSSA missions in the advanced stages of development through the year 2000. Figure 11.2a clearly illustrates the dramatic decline in flight rates that occurred after the end of the Apollo-era, and which in essence persists today.

Since many of the early space science missions served as precursors to the Apollo Moon landings, it is reasonable to ask whether the decline in missions can be explained simply by the completion of the Apollo program. Figure 11.2b is identical to Figure 11.2a, except Apollo precursor science missions have been removed. Clearly, the completion of the Apollo program does *not* alone account for the decline in flight rates. Only 30 of the 96 science missions launched during the 1960s were direct precursors to Apollo. In fact, even excluding the 30 Apollo precursor missions, there were still more science missions launched during the 1960s (66) than there were during the 70s and 80s combined (63).

In both Figures 11.2a and 11.2b one does, however, detect a distinct shift away from programs composed of multiple missions to programs composed of a single mission. This trend becomes apparent immediately after the sharp decline in flight rates in the period 1967-1973, and becomes virtually pervasive after 1980. We note that the hiatus in spaceflight after the loss of *Challenger* in 1986 did not substantially affect the trend toward programs composed of a single mission. Although missions were delayed, no program was cut from multiple missions to a single mission and no program was augmented from a single mission to multiple missions, as a result of the *Challenger* loss.

Figure 11.3 shows the OSSA flights as a function of scientific discipline and the breakdown of these rates between programs composed of multiple missions and programs with only a single spacecraft. From Figure 11.3, we conclude that the general decline in flight rates is not due to a narrowing of the scientific breadth of the NASA program (i.e., as might be expected if one or more disciplines ceased to fly), but instead is manifest as an across-the-board reduction in flight rates for all disciplines. Only unmanned life sciences missions have been disproportionately curtailed in the post-Apollo OSSA effort, although manned life science missions, such as Skylab and Spacelab, have continued. We also find from Figure 11.3 that the trend toward programs composed of a single mission (as opposed to programs with multiple missions) pervaded all four of the major OSSA disciplines after the post-Apollo budget decline. The decline in resources was distributed across the entire OSSA program.

The Reason

The data presented above clearly demonstrate that the character of the U.S. space-science flight program has undergone a substantial shift toward single-mission flight programs. Why did such a change occur? We searched Congressional testimony, read numerous NASA reports to Congress, and spoke with several key NASA managers from the period 1965 to the present.[3] We found no direct documentation concerning the motivations or strategy behind this shift. The most direct responses we heard were that it "just happened" and that it was "an organic" process that occurred without a grand strategic plan. We were surprised to discover that the shift occurred without any formal discussion or documented studies of the risks and benefits of each strategy.

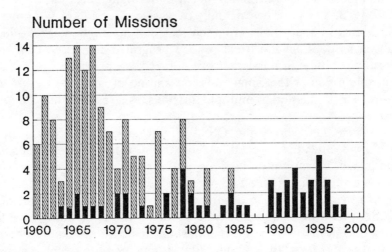

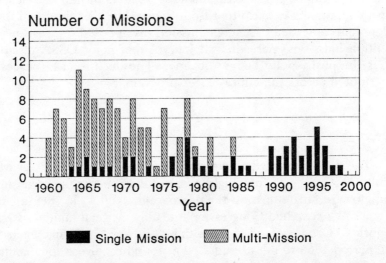

Figures 11.2a and b: NASA Science Missions (1960-2000). The first figure (11.2a) presents the absolute number of space science flight missions launched in each year and the relative proportion of these launches that were part of a program. The second figure (11.2b) presents the same data, but with Apollo precursors removed. Sources: NASA. *TRW Space Log*, V. 23, (1988).

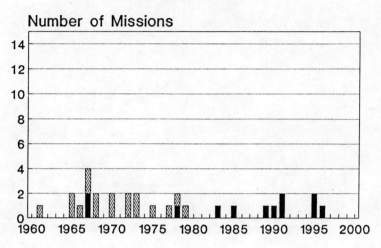

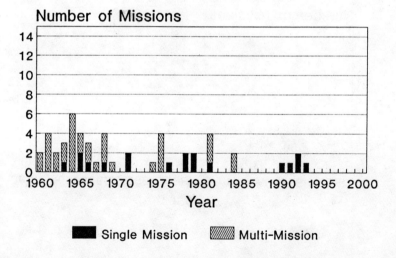

Figures 11.3a and 11.3b: NASA Science Missions (1960-2000). The absolute number of flight missions launched by the Astrophysics (11.3a) and Earth Sciences (11.3b) disciplines in each year and the protion of these launches that were of single mission programs. The Earth Observing System is not included due to program instability. Sources: NASA. *TRW Space Log*, V. 23, (1988).

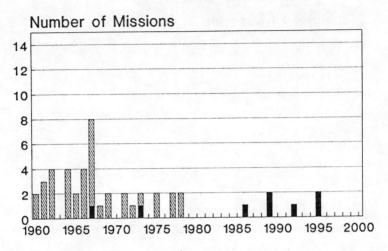

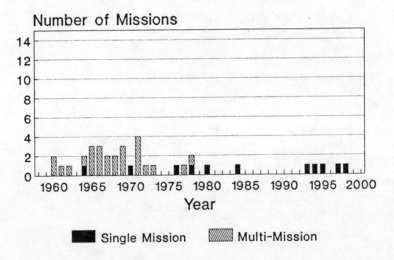

Figures 11.3c and 11.3d: NASA Science Missions (1960-2000). The absolute number of flight missions launched by the Planetary (11.3c) and Solar and Space Physics (11.3d) disciplines in each year and the protion of these launches that were of single mission programs. The Earth Observing System is not included due to program instability. Sources: NASA. *TRW Space Log,* V. 23, (1988).

Considering the 5-7 year lag between program start and the first launch of a mission, we see by returning to Figure 11.2a that the move away from multiple spacecraft missions was made in the late-1960s or early-70s. The poor success rate in NASA's early years created a disposition toward programs with multiple missions. According to Dr. Homer E. Newell, NASA's Associate Administrator for Science between 1967-1973, "[in the early-1960s] NASA had to face the issueof backups for important experiments" [4] and programs containing multiple spacecraft as a result of early experience with the Vanguard and Pioneer programs.

The objective of the Vanguard program was to orbit small scientific satellites, but the program produced only three successes out of 11 launch attempts. The objective of the early Pioneer program was to place a scientific probe near the Moon. This was never achieved, although some of the Pioneer spacecraft did return useful data. Later Pioneer spacecraft were designed to explore interplanetary space and in this task Pioneers 6 through 9 succeeded. The lesson impressed upon program planners was that launching spacecraft was an extremely risky activity and multiple spacecraft within a program was not a luxury, but a necessity if the program was to have a reasonable chance of returning useful data.

In the generous budget climate that prevailed in the early years of the space program, there was also less motivation to accept the increased risk of a program containing only a single mission. The number of investigators interested in space science was relatively small and so there were few constituencies competing for missions. The huge influx of funds into the space program that began in the late 1950s and continued through the early 1960s, as illustrated in Figure 11.4, permitted planners to execute programs containing multiple spacecraft, while still providing missions for each community of investigators.

When NASA's success record improved, however, there was still no direct discussion in the Congressional hearings of making a choice between the two strategies. In 1968 Dr. John Naugle, then the Associate Administrator for Space Science and Applications, noted for Congress the improvement in NASA's success record: "We have come from a period early in the decade when success was a rare and precious commodity, to a time when success is routine and failure the rare occurrence."[5]

Questions about the role of back-ups emerged again when Dr. Noel Hinners, then the Associate Administrator for Space Science, was

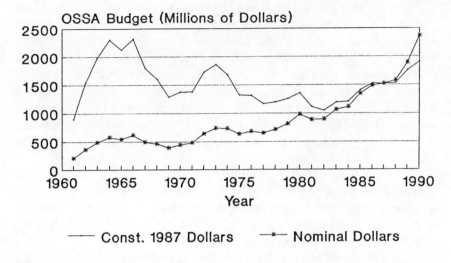

Figure 11.4: NASA Budget History, Office of Space Science and Applications. The NASA budget history in constant and nominal dollars. Source: Konkel, R., *Space Science in the Budget*, Center for Space and Geosciences Policy, Boulder, CO, (1989).

testifying about the 1977 budget request before Congress. Hinners was being questioned about the Viking lander mission to Mars. Two identical spacecraft were launched in this program and a part of each spacecraft was to land autonomously on Mars. Representative Louis Frey (R-FL) was concerned about the risks in the automated landing:

> Mr. FREY. Do you have any backup at all? Have you got two buttons to push? Anything like that?
> Mr. HINNERS. No. Once you commit to a Martian landing the only backup is that spacecraft number two.[6]

By 1976, when Mr. Frey was voicing his concern, NASA had already largely made the shift in strategy to more programs containing a single spacecraft missions and fewer programs with multiple missions.

As NASA budgets dramatically constricted in real terms after Apollo, NASA managers would have faced a declining ability to fund missions while all of the disciplines supported by OSSA created pressure for additional missions. If multi-missions programs were continued some disciplines would likely be hurt. Perhaps the increasing success rate (see Figure 11.5) suggested a different strategy:

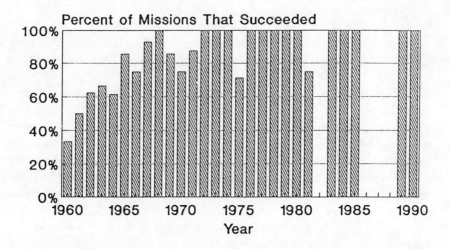

Figure 11.5: Unmanned NASA Science Missions, Success Rate 1960-1990. Note: No missions flown in 1982, 1987, and 1988. No successful missions flown in 1986. Source: NASA.

single mission programs. The effect of single-mission programs was to trade redundancy and depth within each program for the goal of breadth across the entire OSSA effort. This supposition is supported by several arguments.

First among these is the correlation in time between flight rate decline (Figures 11.2 and 11.3) and the space science budget decline (Figure 11.4) that occurred in the post-Apollo era. The decline in resources compelled a choice between pursuing more expensive programs containing multiple missions while neglecting some discipline areas, or pursuing programs composed of a single mission while maintaining overall program breadth.

Second is the fact that in the late-1960s and early-1970s, NASA's early space science successes opened up new disciplines and subdisciplines (e.g., environmental monitoring from space, high energy astrophysics, the exploration of the outer solar system), which increased the number of communities demanding flights from the declining space science budget.

Third, the participants in the program from the 1960s-1970s era with whom we spoke uniformly agreed that in the midst of the post-Apollo budget decline the general perception was that it was a temporary decline which would reverse within 5-10 years. Within the Space Agency it was widely believed that eventually NASA would be rewarded for the successful Apollo missions. The long term decrease in NASA's budgets and flight rates that actually occurred was not expected.

Thus, in aggregate, one imagines that, repeadetly but without specific, NASA-wide strategic direction, individual program managers opted for less-expensive, single-mission flight programs as a way of keeping their disciplines alive and active during a funding decline that was perceived as temporary. The consistently high success rate of the NASA space science program during the late-1960s and early-1970s was likely an important factor in the decision to forgo mission back-ups.

OSSA has fostered a close relationship with a broad range of investigator communities since NASA was founded in 1958. By the end of the Apollo program, these investigator communities had become well entrenched constituencies of OSSA. Rather than losing or offending a particular constituency by informing it that it might have to wait years for new data, or that a mission might never come, management may have found it preferable to accept the increased risk of a program containing a single mission, especially as this risk was decreasing.

The Ramifications

In the context of spacecraft performance and program diversity, we believe that the *de facto* decision for OSSA to remain broad at the expense of redundancy can be argued to be a success. NASA's space science program is broader today than in the early 1970s. New disciplines such as microgravity and earth resources are now being supported by the program. Furthermore, the program today is demonstrably broader than the space science programs of Europe, Japan, or the Soviet Union.[7] Thus, the program has maintained and, in some cases, increased its scope without a single catastrophic science mission failure even in the wake of the post-Apollo budget decline.

Derivative benefits of the single-mission program approach are that it maximizes the design and test experience of space engineers by maximizing the number of unique spacecraft and ground systems that are constructed, and that it increases the amount of space science compared to a program in which fewer programs are funded, each with a series of similar spacecraft.

While we believe that the single-mission program approach has been largely successful, there are also drawbacks. First, despite the success rate OSSA missions have enjoyed over the past 20-25 years, space flight is still a risky endeavour, as illustrated by recent events, such as the flawed mirror aboard Hubble and the current difficulty in unfurling a communications antenna aboard Galileo. A launch or spacecraft failure in a mission such as the future CRAF, Cassini, or AXAF programs, for example, would severely disrupt an entire scientific discipline for a decade or more while a successor mission is studied, advanced through the queue, funded, and prepared. The impact of such a failure would be made all the more severe by the fact that disciplines such as x-ray astronomy, cometary science, and the study of Saturn have each already waited over a decade since their last flight opportunity (the missions AXAF, CRAF, and Cassini are designed to provide data to each of these disciplines respectively).

Therefore, it appears that the decline in flight rate per discipline has made the risk of failure *more severe* than it was around 1970. The realization that a single failure would doom a program containing one flight mission has not been lost on NASA. Indeed, it supports a philosophy of technical risk avoidance and management conservatism that contributes to relentless pressure on mission funds and even further increases the time between missions, which in turn causes increased technical obsolescence of spacecraft and instrument systems.

The infrequency of missions also impairs the ability of each discipline to respond to new technical breakthroughs and scientific discoveries. As discussed by McCray and Stern,[8] the increasing time between missions in each discipline (now equalling a substantial fraction of a scientific career) exacerbates the desire to make an extremely ambitious, do-everything effort out of the few existing missions, further increasing costs. This may be the significant driver behind the progressive increase in the complexity of a typical OSSA mission. Increasing costs due to increased complexity and the enhanced reliability requirement, in turn, contribute to the inability of OSSA to afford more than one spacecraft per mission, driving the situation further from a strategy of multiple missions per program.

Outlook and Conclusions

We have analyzed NASA's record of unmanned space science flight rates from 1960-1991. The objective of this analysis was to assess the degree to which NASA shifted from programs composed of multiple missions to programs composed of only a single mission. In the process of analyzing these data, the pattern of the decline in pace of NASA's space science program was also quantitatively revealed. There were over 90 science missions flown in the 1960s, but there were only 15 flown in the 1980s. This decline cannot be explained by the presence of missions supporting the Apollo program in the 1960s or the loss of Challenger in the 1980s.

As our data show, no particular discipline was singled out to bear the brunt of the decline in missions, rather each of the major scientific disciplines within OSSA (astrophysics, planetary science, earth science, and solar/space physics) has flown approximately the same number of missions during the period 1960-1991. In every discipline except astrophysics, the flight rate has declined precipitously since the late-1960s.

Our data revealed a substantial shift away from programs containing multiple missions to programs composed of only a single mission. While virtually all missions flown in the 1960s were within a multi-mission program, only 2 of 15 missions in the last decade were part of a multi-mission program. This change in character is evident in each of OSSA's scientific discipline Divisions.

Looking into the future at missions which are in the advanced planning stages (but excluding the EOS program, which is presently in a state of flux with an uncertain outcome), we found that virtually all of the unmanned missions presently planned remain single-mission programs. The strategy of single-mission programs has clearly become firmly entrenched in NASA culture, and will persist at least throughout the remainder of the century.

In investigating the reasons for the policy shift from multi-mission to single-mission programs, we could not locate documentation describing this dramatic change. Interviews with former NASA managers, however, allowed us to develop a hypothesis about the shift. We believe that given the combination of budget pressures in the post-Apollo years and NASA's increasing success rate, individual program mangers traded redundancy within a program in favor of retaining the breadth of the entire OSSA program.

Has this strategy been a good one? Gauging by the success rates it seems to have been: only 2 of 38 missions launched since 1976 have failed (1976 is approximately when programs affected by the strategy shift began to be launched). Further, OSSA has successfully retained and even expanded its scientific breadth, despite the budget pressures of the 1970s and 1980s.

A drawback of the single-mission per program strategy, however, has been its implication in the trend of escalating cost and complexity of science missions. This trend is described in detail in McCray and Stern's chapter in this volume. We are unaware of any study attempting to quantify whether the cost escalation in the development of each spacecraft offsets the cost savings of building a single spacecraft. We recommend that such a study be undertaken.

While the single-mission per program strategy pursued by NASA during the last two decades has been successful on its face, its secondary drawbacks have not been sufficiently debated. Congress and the Executive Branch must be prepared to share the blame with NASA for the damage to specific scientific disciplines as a result of future losses, since by their acquiesce to NASA's single-mission per program strategy, they have given tacit approval to this form of programmatic risk taking.[9]

Notes

1. The other planned Great Observatories are the Advanced X-Ray Astronomy Facility (AXAF) and the Space Infrared Telescope Facility (SIRTF). Both are planned for launch around 2000.

2. See, for example, Gary Stix, "Another Small Step; Space Veterans Plan a Low-Budget Moon Mission," *Scientific American*, January 1991.

3. Interviews with Donald Hearth, Noel Hinners, Lennard Fisk, Geoff Briggs, Franklin Martin, Jack Townsend and Robert Smith.

4. Homer E. Newell, *Beyond the Atmosphere; Early Years of Space Science*. National Aeronautics and Space Administration, SP-4211. 1980.

5. U.S. House of Representatives. *Hearings before the Committee on Science and Astronautics [1969 NASA Authorization], 90th CONGRESS, 2nd Session.* February 7 and 8, 1968. Statement of Dr. John E. Naugle, Associate Administrator for Space Science and Applications.

6. U.S. House of Representatives. *Hearings before the Committee on Science and Astronautics [1977 NASA Authorization], 94th CONGRESS, 2nd Session.* February 4, 1976. Statement of Dr. Noel W. Hinners, Associate Administrator for Space Science.

7. Some nations have chosen to pursue a strategy of excellence in a single discipline at the expense of breadth in their programs. The emphasis of Japan's program on X-ray astronomy, Canada's program in polar remote sensing and Sweden's auroral science effort are notable examples.

8. Richard McCray and S. Alan Stern, "NASA's Space Science Program: The Vision And The Reality", this volume. 1992.

9. The authors acknowledge the generous assistance of R. Brunner, R. Byerly, L. Esposito, D. Hearth, R. McCray, R. Oberman, R. Pielke, Jr., and M. Shull in the preparation of this paper.

Chapter 12
EOS, THE EARTH OBSERVING SYSTEM: NASA'S GLOBAL CHANGE RESEARCH MISSION

Ferris Webster

I WILL DO SUCH THINGS,
WHAT THEY ARE YET, I KNOW NOT;
BUT THEY SHALL BE
THE TERRORS OF THE EARTH.
—SHAKESPEARE

Introduction

There is worldwide concern over the economic and social implications of global environmental changes, both natural and human-induced. We recall recent summer droughts in the United States. We hear with concern about the Antarctic ozone hole and its link to man-made chlorofluorocarbons. We witness a widespread debate over mankind's role in the greenhouse effect. Leaders in a number of countries have called for a global program of research and action to understand and reduce the effects of human-induced environmental change. The United States government has been criticized for not acting vigorously enough to minimize those effects. At the same time, many in the scientific community feel that we do not yet understand the earth system well enough to be able to take effective action.

In the face of scientific uncertainty and international concern, the U.S. has proposed a major scientific research program. The U.S. Global Change Research Program (USGCRP) is a key component of President Bush's approach to maintaining U.S. leadership in protecting the global environment. The Program is intended to reduce scientific uncertainty and develop the means for scientific predictions. With scientific progress should come the basis for sound policy responses.

NASA has responded to the challenge of understanding global change with a program called *Mission to Planet Earth*. It has two parts.

The larger component is *EOS, the Earth Observing System,* NASA's major contribution to the USGCRP. It is intended to acquire and assemble a global database over a fifteen-year period. It is to combine the means for making observations and interpreting data with a scientific research effort. The result is planned to be an integrated approach: a system that provides the geophysical, chemical, and biological information necessary for an intense scientific study of planet Earth. NASA sees EOS as "the cornerstone of a long-term program to document global change."[1] Documenting global change will require both space-based and in-situ measurement programs. EOS, as the cornerstone of the USGCRP, must succeed if the USGCRP is to succeed.

The other component, entitled *Earth Probes*, consists of a number of science missions. Some will be near-term and others will operate in the EOS time frame, for missions that require a different orbit or platform than that offered by EOS. *Earth Probes* provides a collection of environmental sensing missions, generally focussed on observing specific Earth processes. Many of these missions have been in the NASA pipeline for some time.

NASA must be doing something right: it has convinced its other-agency partners and the Office of Management and Budget that it merits the lion's share of the USGCRP. The NASA component alone makes up 64% of the budget proposed for the USGCRP in Fiscal Year 1991. That is, the administration and our nation are devoting the majority of the resources for describing, understanding, and predicting global change to NASA's space-based program. In turn, most of the NASA program is devoted to EOS. EOS is thus a program that merits a close look.

NASA has set lofty and admirable goals for Mission to Planet Earth: "It will provide the necessary comprehensive global observations of Earth which will reveal how the processes that govern global change interact as part of the Earth System. This understanding is critical to the development of models for predicting future environmental change. With a continuing, comprehensive data set from EOS, it will be possible to update and enhance the models so that they can provide the vital information needed about environmental change on local, regional, and global scales."[2]

EOS is more than a satellite: it consists of a space-based observing system, a data and information system (EOSDIS), and a scientific research program.

The Space-Based System

The EOS observing system is intended to be comprehensive and global, to cover the earth over broad and high-resolution spatial scales and over long time scales. The program will consist of two series of polar-orbiting platforms. The first platform of the first series is expected to be launched in 1998.

NASA plans call for a long time-series of measurements to obtain a 15-year record of observations using a series of identical satellites, each with a 5-year lifetime. An EOS-A series will alternate with an EOS-B series to assure continuous measurements in two parallel series of instruments. The duration should encompass a large span of major environmental change, including several atmospheric biennial oscillations, three to five El Nino events, and an entire solar cycle.

The EOS platform will be large, accommodating suites of sensors so that multiple views of the same location on Earth will avoid atmospheric-induced uncertainties. EOS will take coordinated simultaneous measurements of the interactions of the atmosphere, ocean, solid earth, and hydrologic and biogeochemical cycles. At least a large part of the payload is supposed to be fixed for the duration of the program to give a consistent set of observations.

The related NASA initiative, Earth Probes, will examine specific Earth processes using smaller platforms or different orbits from EOS. Earth Probes will begin in the early 1990s, before the first EOS platform is launched and will thereby provide some critical near-term observations.

EOS will be complemented by European and Japanese platforms as well as ongoing National Oceanographic and Atmospheric Agency (NOAA) and Department of Defense operational environmental satellites.

With most NASA flight programs, about seventy percent of the funding is allocated to the spacecraft hardware, with the remainder supporting ground-based activities. With EOS, because of the science program needs, about 60 percent of the funding is planned for the ground-based activities, including research and the EOS Data and Information System.

The EOS Data and Information System

NASA plans propose that about half of the EOS budget be devoted to the ground-based component, the largest part of which is the EOS Data and Information System (EOSDIS). Allocating such a large proportion of the budget to the data system component is unprecedented for a NASA science mission.

The amount of scientific data collected by EOS sensors will exceed anything in NASA's experience. EOSDIS may be managing more than 10 terabytes of data per day. To cope with this flow, the scope of EOSDIS will exceed any existing civilian environmental data management system. NASA plans call for EOSDIS to be in operation before the launch of EOS.

EOSDIS will be more than a system to manage data. NASA has lumped a number of functions into EOSDIS that extend it beyond the classic definition of data management:

Mission planning, scheduling, and control
Instrument planning, scheduling, and control
Resource management
Communications (networking, command, and control)
Computational facilities (for processing EOS data)
Data product preparation
Archiving and distribution of data and research results

Normally little more that the final bullet in the above list is what is meant by "data management". The management and operational functions extend the role of EOSDIS. EOSDIS is still evolving and the plans for its functions and architecture are likely undergo modifications.

In support of science, EOSDIS should provide access to data from EOS, from other satellite systems, and from ground-based measurements. It will support the research and analysis of existing datasets and the pre-EOS mission data (e.g., Earth Probes). EOSDIS is planned to acquire and manage a comprehensive, global, 15-year dataset. To do this effectively, it must maximize the scientific usefulness of the dataset and assure easy access by the research community.

EOSDIS design objectives call for a system that is accessible to researchers. It will make available all science data products, models, algorithms, and documentation for the EOS mission. Its policy will be full and open sharing of the full suite of global data for all global change researchers. It will be a distributed system that takes advantage of scientific expertise. Such a system goes well beyond anything that has been done in this area to date.

As far as comprehensive data and information management goes, the current U.S. capability is in shambles. National data centers have been poorly treated during decades of budget scrimping. While there are excellent governmental, university, and research-program-based data centers, there is little experience of integrating components into a coherent system. Global change research may be the catalyst needed to stimulate the improvement and effective operation of existing separate elements. To meet EOS needs, they then must be integrated into an effective national data and information system. The promise is there, and EOSDIS will be challenged to succeed.

EOS Science

The EOS science program is addressed to what NASA calls Earth System Science. This is a new concept, focused on describing the Earth system components, how they interact, how they have evolved, how they function, and how they may be expected to continue to evolve. The goal of Earth system science is to develop the capability to describe and predict environmental changes, both natural and human-induced.[3]

The EOS Science program has seven interdisciplinary themes:

Climate and radiation balance
Oceanic and atmospheric circulations
Global hydrologic cycle
Biogeochemical cycles
Ecosystem dynamics
Atmospheric chemistry
Geological and geophysical processes

These themes are not the same as the priority areas of research identified by the USGCRP planners. (See the next subsection for a list of the USGCRP Focused Program Areas.) Rather they emphasize areas which are seen to be those in which NASA can make the greatest contribution.

The U.S. Global Change Research Program is an unprecedented attempt to understand the global environment. The program aims to develop understanding to help in predicting the consequences of man-made and natural events. Even more challenging is the need to develop policies to mitigate or adapt to global change. To be successful, the research program must focus on the scientific objectives and evolve to meet them. Because our current understanding is weak, the global change research program will have to adapt flexibly to a changing knowledge base. That will be a tall order in a multi-agency program where the natural tendency of the managers running it will be to assure maximum stability for their own component.

EOS science plans tend to be focused on exploiting the measurements from the EOS hardware. It will be critically important to coordinate the space-based results from EOS with non-NASA surface-based and atmospheric-based research. EOS measurements alone will likely not provide the information needed for understanding the global environment. Scientific coordination with non-NASA spacecraft, from other U.S. agencies and from other countries has been poorly defined to date. This may be due in part to the mismatch between NASA's 15-year planning scale for EOS and the year-to-year planning that is typical of most agencies.

The Lion's Share

The EOS program is a major undertaking. NASA planners talk about the expenditure of $17 billion in 1990 dollars through the year 2000. That is likely only about a third to a half of what the total expense will be, since a 15-year dataset will not be in hand before the year 2013. Though the total mission costs are unclear, it is clear that NASA will be challenged to obtain the budget levels necessary to implement the program that is now proposed.

The Bush administration has proposed a USGCRP budget of $1,034 million for FY 1991.[4] Of that amount, $661 million are proposed for NASA's Mission to Planet Earth, that is for Earth Probes and EOS. Thus NASA's program will account for more than 60% of the total USGCRP budget, at least at the beginning of the program.

It may be interesting to look at the breakdown of the proposed 1991 USGCRP budget by Focused Program Area. The budget figures by program area are listed in Table 12.1 in descending order of priority.

Table 12.1. Proposed 1991 USGCRP Budget by Focused Program Area.

Focused Program Area	NASA budget	Total U.S. budget	NASA percent
Climate and Hydrologic Systems	303	462	66
Biogeochemical Dynamics	98	266	74
Ecological Systems and Dynamics	90	179	50
Earth System History	0	19	0
Human Interactions	0	15	0
Solid Earth Processes	63	81	78
Solar Influences	<u>07</u>	<u>13</u>	<u>54</u>
Total Program	661	1034	64

Note: The NASA and total FY 1991 budget figures are given in millions of dollars.

Source: Committee on Earth Sciences (1990).

Many Questions

In a pioneering undertaking of this kind, many questions arise. In the remainder of this paper I'll raise some of those questions, even though (or because) I don't have the answers. I hope this will serve as the basis for beginning a debate on these and other related questions. I put emphasis here on EOSDIS, since that is the program component that seems most original.

The EOS role in the USGCRP

With the EOS program as its principal instrument, NASA has been successful in obtaining the largest role of any agency participating in the USGCRP. It is not hard to see why EOS is such an attractive program. EOS is global in scope. With a varied suite of sensors, it can address many multidisciplinary scientific questions. Because it is technically based, its characteristics lend themselves to detailed advanced planning. Even though that may be counter to the need for scientific flexibility as the program evolves, the air of specificity is attractive to program planners. Such a big technical program is attractive to the influential aerospace lobby. It doesn't necessitate extensive negotiations with other countries. EOS doesn't deal with politically unattractive issues such as human interactions with the environment. It seems to offer the possibility of high-technology scientific solutions which will help solve the political, social, and economic issues that arise from global change.

EOS as planned avoids the controversial social issues associated with global change. But is that justified when the USGCRP is directed to providing the base for policy decisions? Isn't the human-interaction part of global environmental change the driving force? NASA's program devotes nothing to the Human Interactions component of the USGCRP. Nor do the plans for EOSDIS try to relate the other scientific measurements to the objectives of the Human Interactions program. The absence of any planning on the human side may be indicative of the biggest gap between global change research needs and NASA's program.

The issue of global change is one that is politically difficult. President Bush and his advisors have received considerable criticism from the environmental community for what is seen by some as a failure to act decisively in the face of environmental degradation. On the other hand, the issues are confused by scientific uncertainty and confounded by the different pulls of many strong, diverse interests. Thus, it is natural to seek solutions which resolve the scientific issues but present little political difficulty.

The EOS program seems an ideal choice. This high-technology program looks at Earth remotely. Finally, it is the kind of program that can be controlled by the U.S. While international collaboration is certainly desirable, no particular, specific cooperative activity is ultimately essential in order to obtain a global remotely sensed dataset. (That is, if for example some *in situ* data is needed to calibrate space measurements, there will be a variety of sites available worldwide.) The current administration can look forward to a fifteen-year time-series of global measurements with the possibility of resolving scientific uncertainty!

Will EOS deliver on those promises? I'm not sure, but there are a number of reasons to question the contribution of EOS to the

USGCRP. EOS planning at this stage is voluminous, but is mostly focused on the space-borne instrument suite. For example, it's difficult to find specifics about what is needed in the way of in-situ data. Remotely sensed measurements alone will not suffice. Who will obtain the terrestrial, oceanic, and atmospheric in-situ measurements which will be needed? How will they be entered into EOSDIS? (They will, won't they?) And an even more difficult question: what procedures will be used to merge the trillions of bits of EOS remotely sensed data with the sparse terrestrial datasets? Details of these non-space-based issues are poorly resolved in EOS planning documents.

The lack of a strategy for a complementary in-situ measurement program is probably not NASA's fault. All the agencies involved in the USGCRP share in the responsibility for creating a comprehensive scientific plan.

Similarly, there is as yet no EOS international plan. How will the EOS results be used in collaboration with those from satellite programs and in-situ research of other countries?

Is EOS too big?

EOS is proposed as a multi-sensor mission. Two specific high-priority research areas (the role of clouds and fluxes of trace gases) call for simultaneous measurements from a suite of instruments. This leads to the requirement for simultaneous launch of those instruments on at least one large platform (EOS-A). Whether the second series (EOS-B) will also require simultaneity of measurements is yet to be determined. Because of the size of the EOS-A platform, the only launch vehicle with sufficient capacity to launch it is probably the Titan-IV.

There is considerable debate as to whether such a big payload is really needed. The size of the platform alone is drawing criticism, including in the press.[5] Is simultaneity of measurements from a single platform really such a severe scientific constraint? Some critics say that the program would be less expensive and more flexible if it were carried out with a set of smaller satellites with smaller payloads. The question of payload size is beyond the scope of this paper. At least this issue is being debated. There are other issues which perhaps are more worrisome. Some of those other potential problems have not yet been extensively debated, and warrant discussion.

EOSDIS

EOSDIS is planned as a system to integrate and synthesize information from a variety of sources. Because of the high measurements rates, the capacity of the system will exceed anything of its type done to date. Certainly as planned it will be the largest civilian data system ever created. NASA plans for EOSDIS are not consistent or clear on the point, but it appears that EOSDIS will be the only data system to serve all components of *Mission to Planet Earth*–that is, both the Earth Probes program and the EOS program.

EOSDIS will have to build on the experience gained in some NASA centers, in data centers operated by other agencies, and in some research-group data centers. Overall, however, there is nothing one can point to as an example of what is desired. In fact, there is no model for managing and disseminating large multi-disciplinary scientific datasets for coherent analysis and research. The successful creation of EOSDIS will call for significant advances in data and information management techniques.

NASA deserves credit for recognizing the importance of data management in its plans for EOS. Too many times in the past, missions were flown, measurements were made, but little if any of the budget remained after launch to deal with the data.[6] This time, NASA must be congratulated for including data and information management from the start. This is a good start, but I can't help but express concern that space-hardware overruns must not eat into EOSDIS. Past performance justifies that concern.

Evolution

The management of a very large environmental database is a research problem. EOS will be large and complex. To serve it, EOSDIS must be a system unlike any other yet created. The current understanding of how to achieve such a system is not well developed. Thus the evolution of EOSDIS must take place as a research activity.

The preferred technique for creating EOSDIS is not to build to specifications, but to carry out a data systems research program with the development of prototype components. Those prototypes must be developed in close collaboration with global change scientists. Only when the prototypes have been shown to be effective in meeting the needs of the system should decisions be taken about the initial operational form of the system. The ultimate form of the system will depend on experience with the initial operation of the system. An evolutionary procedure should be followed to ensure that EOSDIS meets global research science needs before the die is cast on the system configuration.

An evolutionary EOSDIS runs counter to NASA instincts. The preferred mode of operation for acquiring a large system is to develop the specifications for it, then award a contract to implement it. This is in fact, the track that NASA managers have been on for some time. However, during the past year criticisms of this approach from the scientific community have been strong.[7] NASA managers have been responding to those criticisms and the situation may be changing. A degree of evolution seems to be entering into EOSDIS development plans. However, I believe that the global change research community must keep the pressure on NASA.

Science Responsiveness

Will EOSDIS be able to respond to the science objectives of the USGCRP? The system plans at the moment seem focused on data and information from the EOS platform. That is a worthy aim, but the scientific problems that EOS is addressing demand a data and information system which can cope with a wide spectrum of global-change data types. A result-oriented EOS program is called for. To meet that aim, EOSDIS must cope with all relevant scientific data and information, whatever its source. EOSDIS cannot be limited to EOS satellite data.

To meet its stated objectives, EOSDIS will have to be extended to include ancillary datasets, such as those held by NOAA and USGS. If it is to serve all the data and information needs of *Mission to Planet Earth*, it must be able to cope with areas of research that are far from NASA's current interest. An example that comes to mind is Human Interactions: EOSDIS has no plans to cope at the moment.

I get the feeling that the EOSDIS managers, who already have their hands more than full, would prefer to limit the system to dealing with the immense datasets from the instruments in the EOS observing program. That's an admirable and challenging goal in itself. However, it's not enough for a program that bills itself as the cornerstone of the USGCRP.[8]

To provide the answers to the challenging problems of the USGCRP, an integrated, multi-disciplinary data and information system will be essential. The country is unlikely to be able to afford a multiplicity of systems. EOSDIS is the obvious candidate for the role. However, if that were to come about a number of issues must be resolved. Can NASA do an effective job simply with the EOS/Earth Probes datasets? If NASA can't integrate other types of data and information, what other agency can? To what extent will other agencies be willing to see NASA play the lead role? Can the federal agencies get their act together?

Working with EOSDIS will likely not be a question of choice: We must have a global dataset to deal with the global change issues, and Mission to Planet Earth is the likely source for that dataset. Thus, even social scientists, who may not like the thought of tying themselves to EOSDIS, may not have the choice. The nature of the scientific problem will force them to work with the system.

A Distributed System

EOSDIS is intended to be a distributed system.[9] That is, there will not be a single central data and information center. Rather, specialized data centers will be located at a number of locations. Those centers are called Distributed Active Archive Centers (DAACs). The system, with high-speed network links, will appear to a user as a single entity. The technology is available to create such a system nationally or even internationally, and I believe NASA will be successful with the technical aspects.

The DAAC at the Goddard Space Flight Center will likely be by far the largest, according to current plans. In addition to DAACs, the EOSDIS plans call for the designation of a number of Affiliated DAACs. These complement the DAACs and provide access to non-EOS datasets and services.

A troubling question is that of how the DAACs and the Affiliated DAACs were chosen. Most of the DAACs (5 of 7) are at NASA Centers.[10] This seems reasonable, since most of these centers have experience working with data of the types proposed in EOSDIS. Nevertheless, as EOSDIS plans evolved the number of NASA centers involved as DAACS increased. Did the Centers argue with NASA management for their piece of the action? Is there reason to suspect a spoils system amongst the NASA Centers?

The non-NASA affiliated DAACs present a slightly different situation.[11] There seems to be no clear commitment to their support by NASA (EOSDIS planning documents are ambiguous on this point). Furthermore, the procedures by which the Affiliated DAACs have been or will be selected are unclear. No peer-review or other external review system seems to have been used to select the non-NASA Affiliated DAACs. NASA will likely face increasing pressure to make its choices on scientific competence and to explain its guidelines. It is not always clear what criteria are being used.

Another unresolved question in the distributed EOSDIS is the role of the other federal agencies in EOSDIS. NOAA operates a network of national data centers and holds the only comprehensive collection of oceanic, geophysical, and atmospheric data. The U.S. Geological Survey (USGS) is the logical source for a lot of terrestrial data. Though current EOSDIS plans toss a nod in the direction of these agencies, there is yet no clear indication that they will indeed play an effective role in EOSDIS. Particularly in the case of NOAA, the requisite support to allow them to contribute effectively has not yet been committed. If the other agencies can't meet the needs of the EOSDIS, how will the necessary in-situ data be handled? Might NASA be obliged to create a duplicate set of in-situ earth-ocean-atmosphere data centers?

Can EOS Evolve?

EOS must evolve in response to changing USGCRP needs. Will it be able to? At this stage, the USGCRP research needs are not clear. The multi-agency group that is developing the scientific plans has not yet been able to specify the details of what is needed for many parts of the program. Though NASA cannot be faulted for not responding to plans that are not yet complete, it must adopt an approach to EOS that is evolutionary and flexible.

The detailed scientific plans that do exist for the USGCRP will likely be strongly modified as scientific ideas about global change evolve. Thus the USGCRP is likely to evolve over the life of the program. Furthermore, individuals will change, technology will progress, and scientific knowledge will evolve. The goals of global change research will change as more is learned about the Earth system. Parts of the USGCRP which today are nearly undefined (like the Human Interactions Program Area) may come into focus.

Will EOS be able to cope with changing research objectives? EOS right now is driven by the set of instruments that NASA is planning to install on the space platforms. The scientific planning is very much oriented around those instruments. As global change scientific planning matures, will NASA be able to respond with a modified EOS program? At this time, there is not even a contingency plan for modifying EOS. There is no plan for incorporating new understanding, new concepts, or new technology. Surely evolution will be needed, but no provisions have been made for it.

A flexible approach to EOS will be expensive. Realistically, the program is likely to operate on a fixed budget. Each time something new is added to respond to the latest scientific challenge, something else will have to be dropped. EOS is already an expensive federal budget item. Keeping options open is a costly way to proceed. The realities of the budget will likely force early decisions about the selection of options.

The problem of EOS evolution is complicated by the requirement for continuity of measurements. That is, EOS must stay flexible, but somehow maintain a continuous suite of measurements. The stated requirement is for at least a fifteen-year series of regular global measurements. Can a balance be struck between a fifteen-year no-gap set of measurements and the need to evolve the program in the face of changing scientific perceptions?

Planning

As has already been noted, EOS is short on scientific planning and documentation. Most of the plans are oriented around the instrument packages to be flown on the mission. There is little to show how EOS will deal with non-NASA satellite data, with in-situ measurements, or with the international global change research program. I found little in the way of formal documents that I could cite for this paper. Much of my source material consisted of copies of viewgraphs from briefings. Why is there so little documentation and planning on what are important issues?

My feeling is that NASA managers just don't have the time to plan, document, modify, or evolve the program. They are already overcommitted and up to their necks in alligators, just keeping up with the intense pace the existing program demands. That pace shows no sign of slowing and without some fundamental changes, I don't see much sign of extensive new planning or documentation in the near future.

The pressure of having to make many decisions synchronously will likely swamp NASA managers, who will pass them to the EOS project, at Goddard Space Flight Center. They in turn, may pass the decisions on to the EOS contractors. This begins to have a familiar ring: the lessons of the Hubble Space Telescope and the Challenger Space Shuttle mission should not have to be re-learned.

The Outlook

EOS planners have not done all their homework. Because of the magnitude of the task that faces them, there's no indication that they will. They see it as sufficient to document the scientific arguments for the EOS instrument payload and to manage that part of the program effectively. Because of the interdisciplinary nature of global change problems, the total job must be a team effort. NASA planners are looking to others to share in defining and building the total USGCRP. All the agencies are responsible for planning the USGCRP. Other agencies are more involved in collecting and analyzing in-situ measurements. Other agencies, particularly the National Science Foundation, are more able to call upon the requisite scientific expertise. Not only other agencies, but other countries are responsible for the international coordination.

What's stopping the other agencies from being more assertive? The problem is that NASA is the 600-pound gorilla in the USGCRP. If NASA does not take the lead in addressing the problem of formulating a comprehensive inter-agency and international program for remote sensing in support of global change research, it is unlikely to get done in time to impact the planning for EOS. Until it is done, it is difficult to argue that the current plan for EOS represents the most cost-effective way of supporting the objectives of the USGCRP.

Notes

1. Goddard Space Flight Center, 1990. *Earth Observing System, 1990 Reference Handbook*, Greenbelt, MD.

2. EOS Program Office, NASA, 1990. *EOS A Mission to Planet Earth*. NASA, Washington, DC.

3. The scientific planning that led to EOS is summarized in: Earth System Sciences, Committee, NASA Advisory Council, 1988. *Earth System Science, A Closer View*, University Corporation for Atmospheric Research, Boulder, CO.

4. Committee on Earth Sciences, 1990. *Our Changing Planet: The FY 1991 U.S. Global Change Research Program*. USGS, Reston, VA.

5. See, for example: William K. Stevens, 1990. "Huge space platforms seen as distorting studies of Earth", *New York Times*, Tuesday, 19 June, 1990, page C1. Eliot Marshall, 1989. "Bringing NASA down to Earth: A $15- to $30-billion earth observing program for the 1990s draws fire for spending too much on hardware, too little on science". *Science*, Vol, 244, 16 June 1989,pp. 1248-1251.

6. For example, Kneale, Dennis, 1988. "Into the void: What becomes of data sent back from space? Not a lot, as a rule." *Wall Street Journal*, 12 Jan 1988, Vol CXVIII, No. 7, p. 1.

7. Science Advisory Panel for EOS Data and Information, 1989. *Initial Scientific Assessment of the EOS Data and Information System (EOSDIS)*. EOS-89-1, Goddard Space Flight Center.

8. Goddard Space Flight Center, 1990. *Earth Observing System, 1990 Reference Handbook*, Greenbelt, MD, p. 3.

9. *Op. cit.*, p. 9

10. The Centers are: Goddard Space Flight Center for climate, ocean biology, and the upper atmosphere; Jet Propulsion Laboratory for physical oceanography and air-sea interactions; Langley Research Center for radiation budget and aerosols; Alaska SAR Facility for sea ice; Marshall Space Flight Center for the hydrologic cycle; EROS Data Center for land processes; National Snow and Ice Data Center for snow and ice.

11. The Affiliated DAACs identified to date are: the National Center for Atmospheric Research; the University of Wisconsin Space Science and Engineering Center; and CIESIN, a consortium of the Saginaw Valley State College, the University of Michigan, and the Environmental Research Institute of Michigan.

Chapter 13
THE FUTURE OF THE SPACE STATION PROGRAM*

Ronald D. Brunner, Radford Byerly, Jr., and Roger A. Pielke, Jr.[1]

AND WHAT ROUGH BEAST, ITS HOUR COME ROUND AT LAST,
SLOUCHES TOWARDS BETHLEHEM TO BE BORN?
—YEATS

Introduction

The Space Station program has become increasingly controversial since its inception in 1984. Two news reports illustrate the polarization of the issue in the summer of 1990. On July 9, the *Wall Street Journal* reported that Congress is threatening to cut off funds in December for NASA's proposed $37 billion space station because it's overweight, underpowered, hasn't proved it can perform important scientific experiments, and may require extra space shuttle flights for assembly in orbit.[2]

On August 20, *Space Station News* reported that sixty-four senators had sent a letter to Barbara Mikulski, head of the Senate appropriations subcommittee, urging full funding for the Space Station program and "understanding" with regard to its weight, power, and external maintenance problems.[3] This is not the first year that problems in the Space Station program have reinforced the program's

---

* Editor's Note: This chapter was completed (aside from minor editorial changes) in September, 1990. It was deliberately not updated in order to facilitate the use of subsequent experience to assess the analysis.

opposition and forced the program's supporters to step up their defense.

This chapter considers how the growing controversy over the Space Station program might be resolved, given a range of scenarios.[4] The first section contends that achievement of the current program baseline is an unlikely scenario, and presents continued program instability and possible program termination as two more likely scenarios. The second section contends that "business as usual" – that is, the *de facto* policy which attempts to stabilize the program and to minimize the appearance of instability – probably leads either to a space station built down in increments or to program termination and possibly a restructuring of NASA. The third section considers how the present Station might be redesigned in order to salvage something from an investment that will approach $4 billion by the end of FY 1990.

This chapter does *not* attempt to predict the future of the Space Station program, nor does it recommend how the growing controversy should be resolved. It does attempt to evaluate testimony on the future of the program in the light of the program's history and underlying dynamics. The purpose is to provide a more realistic frame of reference that the public and public officials might use to make more informed decisions and to reassess the program as events unfold.

Scenarios

Achievement of the Baseline

Early each calendar year, with the beginning of Congressional hearings on authorizations and appropriations for the next fiscal year, NASA officials consistently testify that the Space Station program has been stabilized or soon will be.[5] Stabilization is understood to mean the fixing of the program's design and capabilities, schedule of major milestones, and total development cost, as specified in the current program baseline.[6] Just as consistently, the current baseline turns out to be an unrealistic scenario for the program's near-term future. So far at least, program stability is an annual expectation and an annual disappointment.

Consider a statement by Andrew J. Stofan, then Associate Administrator for Space Station, prepared for hearings early in 1988 on the FY 1989 authorizations:

> Today, the cost of that [current baseline] configuration is identical to what it was when we came out of that review in January 1987. So we have now been a year with this configuration, and the cost has stayed stable for that period of time.... We have not changed the configuration or a system or a subsystem on that Space Station; it is exactly the same as it was.[7]

The implication that the program had stabilized was unwarranted: Within a month of Stofan's assurance, the development plan submitted to Congress showed major program milestones slipped by about one year. The covering letter from Administrator James C. Fletcher, dated April 7, 1988, notes that the plan "outlines a new program schedule. This schedule reflects the revised budget estimates for FY 1988 and FY 1989 and shows the first launch of a Space Station element to be early in 1995."[8]

Consider a statement the next year by James B. Odom, Stofan's successor as Associate Administrator for Space Station, which was prepared for hearings early in 1989 on the FY 1990 authorizations:

> We are still on schedule for all major program milestones... Finally, I want to note that our estimate of total runout costs for the Freedom development effort remains at $13 billion in 1984 dollars. In summary, the program has stabilized. The configuration has been set since 1987, and our cost estimates and schedules remain the same.[9]

Within a few months, NASA initiated the Langley exercise to rephase the program under new leadership. This was motivated by an expected cut of $400 million from the FY 1990 request for the Space Station. But it was also motivated by the new leadership's risk assessment within the program, independent of the budget cut, according to testimony by Odom's successor, William B. Lenoir.[10] Another motivation and a goal of the Langley exercise, according to the new Administrator, Admiral Richard H. Truly, was "to stabilize the program to the degree possible and carry it forward into the design phase."[11]

Finally, consider a statement by Lenoir, now Associate Administrator for Space Flight (which now includes Space Station and the Shuttle), prepared for hearings early in 1990 on the FY 1991 authorizations. The four points emphasized in the closing summary and conclusion of his statement are reproduced here (with emphasis added):

> First, *our objectives for Space Station Freedom have not changed.* For our science program, Freedom offers a laboratory for research... and a platform for instruments.... For technology development, Freedom will be a unique national

facility - a testbed for evaluating technologies, procedures, and design approaches.... For our exploration program, Freedom's manned base represents the establishment of a permanently manned outpost in space and a stepping stone for the missions of the future.

Second, our progress to date on the Space Station has been substantial. *The First Element Launch milestone is one year closer now* than it was at our last budget hearing before this Subcommittee, and our design and development effort is moving forward accordingly on all fronts....

Third, some program adjustments have been required in the last year, but *the basic integrity of the Space Station program is intact.* The reduced funding for FY 1990 and the resulting need to minimize near-term program costs, technical risks, and schedule risks forced some changes in program plans, primarily involving the schedule for outyear assembly milestones and addition of planned capabilities. Despite these recent adjustments, however, *the fundamental objectives of the Space Station program are unchanged, and the planned capabilities of the Space Station are maintained.*

Finally, as you well know, full funding for FY 1991-1993 is necessary if we are to avoid additional disruptions in the Space Station development effort. *The stability that is required can only be provided by multiyear funding....* Renewal of [the current] multiyear authorization and initiation of a multiyear appropriation are essential *if we are to complete the baseline Space Station before the start of the twenty-first century.*[12]

Lenoir's statement is rather cautious in comparison to previous years, but otherwise typical in several respects. First, the general, long-term objectives used to justify the program are featured prominently up front. These objectives do not justify any particular program or budget; they are generally consistent with many different program baselines. Second, the appearance of instability is minimized through a selective review of the program's recent history. Lenoir draws attention to the stability of the FEL milestone, the basic integrity of the program, fundamental program objectives, and planned capabilities, despite recent adjustments. Those recent adjustments — the rephasing that resulted from the 1989 Langley exercise — were not considered minimal by the full House authorization committee four months earlier.[13]

Third, program stability is considered both a feasible and the preferred scenario in this statement. Lenoir describes multi-year funding as a necessary condition for program stability, and program stability is required to complete the current baseline Station before the next century.

By mid-1990, within a few months of Lenoir's statement, a NASA internal assessment of the program acknowledged further instability. According to reports in the trade press, additional schedule slips and content cuts would result from a $195 million cut by the House of Representatives from the program's request for FY 1991. But program officials had already begun to consider another rephasing exercise long before the budget cut because of "major hangups" within the program. These include "a shortage of volume, excess weight, and oversubscription of power requirements, as well as the ongoing maintenance problems and subsequent Extra-Vehicular Activity (EVA) requirements."[14]

In short, achievement of the current baseline program has not been a realistic scenario, nor is it likely to be realistic given NASA's own premises about the requirements for stability. First, NASA leaders have insisted that full funding of the annual request is necessary for program stability.[15] Second, NASA leaders have acknowledged that some major changes, sufficient to destabilize the program, can be forced by internal reassessments that are independent of budget cuts.[16] (To characterize such changes as part of the "normal" process of design is misleading because they have essentially the same destabilizing impact as budget cuts; the normal process accommodates changes through program reserves without destabilizing the program.) If one accepts these premises, then the current baseline will be a realistic scenario only if the program is fully funded and if all changes through design and development are accommodated by reserves. Neither condition is likely to be realized over the next several years, for reasons summarized in the instability scenarios below.

NASA leaders are apparently reluctant to face the logical implications of their own premises and chronic, severe constraints on the federal budget, even when pressed. In the October 31, 1989 hearings on proposed revisions in the program, Representative Harold Volkmer observed that "next year or the following year — somewhere along the line — we may have budget problems again." Then he asked Administrator Truly the obvious question: "Does that mean we do a rephasing of it each time?" Truly's struggle with the question is illuminating:

> Again, I don't know how to guess. Certainly next year's fiscal constraints on the total budget are going to be severe, as they have been this year, but if we are on a program plan that we f and this is why I believe so strongly that, as I think this committee does, if we could get some stability in year-to-year funding so the program could plan on it, you wouldn't see disruptions.
> So I don't know how to answer since I don't know what the potential disruption would be. Certainly if it was major we would have to f we have to reassess it. We can't continue [sic] to provide capability on a schedule without money. It

is a three-legged stool. But I do believe that we are on a plan with what we achieved in the fiscal year 1990 budget to get an achievable budget level for the Station, and we have management in place now that has a tremendous amount of experience in managing very large, complex programs. We have done a lot of work with our international partners to make sure that they understood where we are, and we are continuing that work. So I feel good about where we are... But beyond that, I don't know how to answer your question.[17]

The Admiral begins and ends with the frank acknowledgement that he does not know how to answer, as if rephasing driven by funding shortfalls is a new problem for which there is no relevant experience. The difficulty in answering is attributed to not knowing "what the potential disruption would be." Truly substitutes a hopeful fantasy ("if we could get some stability in year-to-year funding") for the sobering fact that multi-year funding for the Station has been consistently rejected by Congress. He also finds reason for hope ("I feel good about where we are") in a highly selective assessment of the program's recent history. This is a manifestation of NASA's "success-oriented" culture.

A more realistic and balanced assessment would have recognized chronic funding shortfalls and recent changes driven by events within the program, considered them in the light of expected "severe" constraints on the federal budget and the "three-legged stool" ("We can't...provide capability on a schedule without money"), and drawn the logical conclusion: There is little uncertainty about *whether* the program will be further disrupted, even if there is some uncertainty about *what*, specifically, might force the disruptions.

What sustains presentation of the current baseline as the only scenario for the future of the Space Station program is political necessity. Continued instability jeopardizes the program's survival, as Chairman Roe observed in the October 31, 1989 hearings on proposed revisions in the program:

> If the committee could only look forward to continued replanning, reshaping, broken commitments, changing management philosophies, and international embarrassment, I believe we should leave the Station behind us and begin anew with a more resilient plan for the future in space.[18]

Roe was still troubled about program instability eight months later when he requested an audit of program changes by the General Accounting Office: "The rephasing of the Space Station that occurred last fall...was of real concern to the Committee because it suggested serious instability in the program." "Our hearings on the rephasing did not allay our concern." [19]

Similarly, Subcommittee Chairman Nelson, a committed but concerned supporter of the Space Station program, reviewed the program's history of instability and drew the obvious conclusion:

> So it is time to settle on one design, on one design schedule, and stick to it.... The constant changes are causing our international partners great concern and are making it increasingly difficult and expensive to move into the construction phase.[20]

Administrator Truly, paraphrasing his prepared statement, agreed that "as I look at the history of this program...there is a real and true need to provide stability in this program for coming years..."[21]

Those supporters who contend the program *can* be stabilized because it *must* be stabilized to survive may be relying on faith at best and self-delusions at worst, but they are not relying on their own experience with the program. Selective inattention to their own experience apparently reinforces political necessity in sustaining achievement of the baseline as a scenario.[22]

Instability Scenarios

The instability of the Space Station program is a symptom of an underlying problem, the lack of program resilience in an evolving environment. Resilience is the capacity of a program to perform as promised, despite disruptions external or internal to the program. Resilience is necessary because program performance depends upon funding and other critical factors that NASA cannot entirely control or adequately anticipate.

The Space Station program lacks resilience because it was conceived to be a large-scale, long-term program, interdependent with most other U.S. civilian space programs and with some space programs of other nations. Its large-scale and long-term mean that significant disruptions are relatively numerous, because more critical factors are subject to change over a longer period of time. Interdependence means that a disruption anywhere − in the program itself, in other space programs (e. g., the Shuttle), or in the fiscal and political environment − tends to force many changes and further disruptions elsewhere in the program.[23]

Instability will be forced on the Space Station program by appropriations that continue to fall short of the annual request, by problems that continue to arise in development, and by surprises if not accidents that are likely to occur in operations on the ground and in orbit. There is no reason to believe that the program can avoid further disruption by events of these kinds. The specifics of the disruptive events are uncertain, but the specifics do not matter.

First, the near-term budget situation is not likely to improve. The Administration recently increased the projected federal deficit for FY 1991 to $168.8 billion. This exceeds the target set in the Gramm-Rudman-Hollings deficit reduction act by more than $100 billion. It implies automatic funding cuts late in the fall of 1990 of about 38% for non-defense discretionary programs, including civilian space.[24] Current negotiations to bring the deficit under control are more likely to change the target than to achieve a major substantive agreement on expenditure and revenue decisions. In any case, the deficit is a chronic problem that is widely understood to require much more political courage and will than has been evident.[25] Continued severe constraints on the federal budget are the only plausible projection.

NASA has fared extremely well in the competition heightened by severe constraints on the federal budget, but not well enough to fully fund the Space Station budget request or to secure multi-year appropriations. Chairman Traxler summed up the situation in debate on the FY 1991 House appropriations bill:

> The fact is that the space station program is in deep trouble. And unless we fund this program and get it moving and insist that the problems are solved - we are going to continue to spend billions and never build a single piece of hardware....
>
> Frankly, the fact is that we have probably tilted too much in favor of NASA. What we should have done is to add some extra money over and above what we provided for veterans medical care. But we stuck every last dollar into NASA that we could find....
>
> And if anyone is not happy with that - stay tuned - because it's going to get a lot worse in the budget wars ahead.[26]

The House bill cut $195 million from the President's FY 1991 request for the Space Station. Whatever its size, the Space Station budget cut for FY 1991 is not merely a one-year inconvenience.

What can NASA do to increase the Space Station budget under these circumstances? All the obvious means have already been tried without succeeding in stabilizing the program.[27] However, further deterioration in program performance — cutbacks in capabilities, schedule slips, cost increases, and the like — could be used to bolster the argument that the program needs additional funding to avoid termination and to protect all those who share some responsibility for the program.

Second, problems in the development of the Space Station are increasingly likely to surface as the program continues. These problems will continue to be characterized by NASA officials as "normal." Nevertheless, responses to these problems will disrupt the program even if it achieves full funding. The following unresolved problems illustrate the potential for further disruption:

* Electrical power for scientific utilization of the current baseline may be insufficient (according to earlier NASA estimates) after "housekeeping" functions are provided for.[28]

* The excessive amount of EVA time required for maintenance during assembly will force substantial simplification of the current baseline or other changes.[29]

* The excess weight of current elements — roughly 150,000 pounds over the 512,000 pounds allocated in November, 1989 — is likely to force further reductions in capability and greater dependence on Shuttle.[30]

* Some Space Station assembly elements still do not meet the Shuttle's payload constraints.[31]

* The threat from medium-sized (between 1 and 10 cm) orbital debris is uncertain but acknowledged by NASA to be significant.[32]

* Interactions between plasma in space and direct-current power distribution could cause electrical arcing, adding to crew safety and maintenance problems.[33]

* Knowledge of the many physical and biological factors involved in protecting the crew from radiation is incomplete and uncertain.[34]

Attempts to find satisfactory solutions to such problems, whether successful or not, will take time and money and probably will compromise capabilities.

Third, problems in operations are inevitable. The Shuttle is the Station's Achilles heel in this respect. Assembly of the Station can be disrupted by Shuttle-related events like those which have already occurred: An accident during launch preparations,[35] the discovery of hydrogen leaks in the fuel system,[36] or loss of another Shuttle. The Office of Technology Assessment estimates that the Shuttle is about 98% reliable. If so, we can expect to lose one Shuttle in approximately 34 flights. Using 1989 Shuttle manifest data, that means the probability of having all four orbiters available at the scheduled start of Space Station assembly was only 0.28; the probability of having them all available at the scheduled completion of assembly was only 0.12.[37] Further, station assembly must proceed in the exact sequence planned, and yet Shuttle flights, historically, simply do not fly in the planned order.

Each successive flight in the dozens of flights required during Station assembly presents a difficult and exacting docking problem, because the partially-assembled Station will be somewhat different each time. Damage to or loss of any crucial part could set the program back years, because there are few spare parts. Moreover, mistakes and accidents are relatively likely because the Space Station, unlike most previous space systems, "is never fully assembled and checked out on the ground."[38] Uncertainty about the physics of the plasma problem can be generalized to other aspects of the program: The Station "is too big to test in the environment it will work in. You've got to make the physics up as you go."[39]

Such problems in funding, development, and operations interact: Responses to one problem tend to be constrained by, and to exacerbate, other problems. For example, obsolete underestimates of the threat of orbital debris were retained in April, 1989 because "changed debris environment [estimates] would result in design modifications that NASA could not afford."[40] Similarly, a change request for a positively-grounded power system to address the plasma problem was killed by a quick cost estimate, about a half billion dollars. The plasma only became a problem because the program switched from alternating to direct current in response to earlier problems.[41]

Adding Shuttle flights is one response to the payload-constraint and weight problems, but not a sufficient response. Moreover, adding Shuttle flights would increase the probability of losing a Shuttle and Station components during assembly, increase the amount of EVA time required, increase exposure to unknown radiation and debris hazards, and add significantly to operations costs.

Many interacting problems complicate management and integration, which were already complicated enough. An anonymous official working on the weight problem observed that "while the PDRs [preliminary designs reviews on individual systems] are going on, in the next door or two doors down they're scrubbing the [same] systems."[42] The scrubbed version of a system forces changes in interdependent elements that assumed the previous baseline; and all changes from scrubbing must be coordinated through an unnecessarily complicated management structure involving several NASA centers in major roles.[43] Proposed changes are especially disruptive because they are much more numerous and because they propagate many "what if?" exercises for interdependent elements, diverting time and effort from work on the baseline version.

In summary, the Space Station program will continue to be disrupted, and because the program lacks resilience a disruption anywhere tends to multiply problems elsewhere. Failure to respond to such problems adds to technical risk in the program. But responses put additional pressures on reserves of attention, time, and funds, and such reserves already have proven insufficient to stabilize the program. The result is continued instability, manifest in reduced program capabilities, delayed program milestones, and added program costs. NASA recently described how one problem, a budget cut, compounds other problems:

>...the quick look assessment shows that a $150 million reduction in FY 1991 would increase total program costs by $100-250 million, would delete some content, would delay key operational milestones 6-9 months, and would delay planning for potential growth to meet future requirements by one year. In addition, the launch of the European and Japanese modules would be delayed 7-8 months under the current assembly sequence, with some accompanying cost increases.[44]

Termination Scenarios

Termination of the Space Station program has been raised as an alternative to program continuation for the last several years. NASA spokesmen typically select and emphasize the adverse consequences of termination for others as well as for NASA itself. For example, in the 1989 authorization hearings, James Odom, Associate Administrator for Space Station, testified that:

> A significant cut below where we are, I think, would be tantamount to killing the program. And I think that to do that would be almost irresponsible on our part, especially looking at the commitments that our international partners have made and the investment that our contractors and we have made up to this point in time.[45]

In contrast, threats of termination coming from Congressional critics typically select and emphasize problems within the Space Station program.[46]

Such statements are best interpreted as brinksmanship in the continuing controversy over the program. For NASA, the threat of termination is a tactic to pressure Congress to fund the Space Station program at a level NASA considers to be adequate. For Congressional critics, the threat of termination is a tactic to pressure NASA to improve program performance according to whatever criteria the critics insist upon, including program stabilization. The assumptions that Congress *can* provide funding adequate to support NASA's desired program and that NASA *can* stabilize the program are unexamined and probably unrealistic.

No can predict program termination or continuance with any confidence. The question for present purposes is whether termination of the Space Station will become more or less likely over the next several years, and under what circumstances? The answers depend upon the perspectives of those who would be most directly involved in a termination decision.

For NASA's leadership, it is doubtful that termination would be preferred over program continuance under any plausible circumstances, despite testimony that termination makes sense below some unspecified minimum level of funding. Continuance serves the institutional interest in maintaining employment and the shared vision of human exploration.[47] Termination would be devastating in these terms, and would threaten NASA as we know it because NASA has centered the civilian space program on *a* station or *the* Station over the last two decades. From this perspective, the frustrations of managing and attempting to rejustify an unstable program appear to be rather minor.

Supporters of the program in the Administration and Congress are also frustrated by the program's instability and poor performance, which complicates the problem of selling the Space Station. Selling the program, despite its poor performance, is sometimes the explicit purpose of hearings according to chairmen of the Senate and House authorization subcommittees, respectively:

> Today, we will try and compile a record that will help us argue for [Space Station] as a budget priority in the turbulent days ahead.[48]

> ...we want you to reassure us that [a slip in the schedule] is not the case...so that we can go out and continue to sell to the Congress the appropriate amount of money that is needed for this project which is so enormously important to our country.[49]

NASA's commitment to First Element Launch in March, 1995 attempts to address the concern over schedule slips, even if slips in other milestones are unavoidable. So far at least, the frustrations of selling a troubled program are more than compensated by other factors, including tangible benefits for some constituencies and faith in the program's many general, long-term justifications.

Skeptics and critics of the program tend to be frustrated by the lack of good alternatives. From their perspective, program continuance is a bad investment of public resources, but any attempt to terminate the program would be a major political effort and risk, which, even if successful, would disrupt the entire civilian space program. Moreover, the public would demand to know who was responsible for terminating a program which had already spent nearly $4 billion through FY 1990; and everyone who shares some responsibility would be predisposed to point the finger elsewhere. The resulting waves of recrimination could touch everyone involved with the program in NASA, the Administration, and Congress, and wreak major political damage on some of them. With a few exceptions, members of Congress have been unwilling to take initiatives toward killing the program.[50]

The net result has been a deadlock: NASA and Space Station supporters have been unable to "build it right" and the opposition has been unable to "not build it at all," and so the program continues. But the perspectives behind the deadlock, and the outcome, are subject to change as present trends unfold.

One trend is simply growing momentum. Some believe that the Station program already has enough momentum to carry it "over the hump" to assembly complete. Others believe the program will reach that point with completion of the Critical Design Review, the fabrication of hardware, or First Element Launch. Momentum means that the stakes increase as a function of time: For supporters, each additional appropriation or milestone intensifies the motivation to continue and complete the Station; for opponents, each additional investment more effectively deters termination initiatives, because it becomes increasingly difficult to explain why more money was "wasted" on plans or hardware that will not be used.[51]

Continued program instability works in the opposite direction. Evidence of program instability leaked or reported to the press tends to stimulate and surface other reports of problems through a bandwagon effect.[52] Such reports tend to undermine "soft" support based on nothing more tangible than faith in the program's justifications. As stepped-up selling efforts produce diminishing results, core supporters of the program become demoralized and eventually concerned that their visible support is becoming a personal political liability. Under these circumstances, it is prudent to keep one's head down and look for other causes to champion. Meanwhile, skeptics and critics are encouraged by the increasing results of their efforts and begin to sense a political opportunity. As such dynamics build up the potential for termination some minor event — a blink in the game of brinksmanship, perhaps — can trigger termination.

The probability of termination will increase from year to year, so long as as instability and related problems in the program continue to occur and continue to be reported. Under these circumstances, the program's large-scale and momentum work to its disadvantage: It is perceived as a large and growing mistake, increasingly visible and therefore increasingly costly to support. Moreover, each year the program is stretched out makes it more vulnerable to scientific and technological obsolescence and to competition from any programs, American or foreign, with overlapping objectives.

Some NASA leaders have been especially concerned about slips in the schedule, one form of instability, as a threat to program continuance. Acting Administrator Myers testified that "there is certainly history that says if programs continue to slip, they finally get cancelled."[53] During the 1989 Langley rephasing, Administrator Truly recalled his experience as an astronaut assigned to the manned orbiting laboratory (MOL), an ill-fated Air Force project of the 1960s. According to *Space Station News*:

> That program cost $3 billion and slipped 3 times" before it was cancelled, Truly recalled last week. "The space station is at that point. And I don't want to slip it." Said one contractor: "That experience with MOL really scarred him. This

[Langley exercise] comes right out of that - he believes you must get something into orbit.[54]

In summary, it appears likely that (1) achievement of the current baseline for the Space Station program is as unrealistic as achievement of previous baselines; (2) the program will continue to be destabilized by funding shortfalls and by internal problems arising from program development and operations; and (3) the probability of program termination will increase so long as program instability continues to be reported. What can be done about the Space Station program?

Business As Usual
One policy alternative is business as usual, the *de facto* policy of the Space Station program. This means continuing efforts by NASA and others to stabilize the current baseline design and basic performance criteria, especially capabilities, schedules, and total costs. This also means continuing efforts to avoid the appearance of instability when changes cannot be avoided, in order to reduce the threat of termination.

Business as usual is an unrealistic policy if stabilization of the current baseline is, in fact, the principal objective. As argued above, all previous efforts to stabilize the program have failed, and there is no reason to believe that such efforts will be any more successful over the next several years, even with full funding. Full funding is unlikely, and it will take additional time and money beyond the current baseline to find and coordinate changes in response to the major problems already identified. Thus the descoping of the Space Station to Block I in 1987, the rephasings in 1989 and 1990, and lesser changes in program content are very likely precursors to additional exercises to delete or defer work and to incur the higher costs resulting from such changes.

Under these circumstances, resulting political pressures to minimize the appearance of instability, and to exaggerate claims of stability, would compound the technical risks of business as usual. Lower-level officials would be reluctant to raise additional problems that might jeopardize the program; and higher-level officials would be reluctant to hear about additional problems that might call into question their credibility as well as the program's viability. Such internal problems of communication were major factors in the Challenger accident, according to subsequent investigations.[55] The lesson is that problems left unattended do not disappear. Rather, they tend to show up later in operations where the differences between appearance and reality are reconciled.

Over a series of descoping and rephasing exercises, each of which adds to development and operational costs, business as usual could produce an operational space station. This progeny of the Space Station, several generations removed, would be less than currently promised and much less than was promised in 1984. It would also be plagued by intermittent operational problems, as the Shuttle still is, to the extent that internal problems are left unattended. The likelihood of this outcome depends upon the success of NASA's efforts to minimize the appearance of instability — among things, focusing attention on the general, long-term goals of the program, and encouraging the Administration, Congress, and the public to overlook the history of promises not fulfilled and problems not resolved.[56]

Business as usual could also lead to termination of the Space Station — an outcome which, we have argued, becomes more likely as the series of descoping and rephasing exercises continues. Termination of this centerpiece of the civilian space program, could call into question the entire NASA structure and make restructuring difficult to avoid. Dale Myers testified to that effect when he was Acting Administrator of NASA in 1989:

> I think NASA could be in a tailspin without the Space Station. I think the whole strategy of our program, all the long-range planning that I have been associated with since 1968, has involved the Space Station. And so, the idea of having a [civilian space] program without a Space Station is abhorrent to the highest degree, and I believe that if there were a major cut in the program, we would have to look at the total infrastructure of NASA again. It would take a different NASA to operate without a Space Station.[57]

The many frustrations that have accumulated in and around the civilian space program in the post-Apollo era would be brought to bear on the issue of restructuring, making the outcome highly unpredictable — especially given the lack of prior planning for such a contingency.

On the other hand, termination of the Space Staion would benefit NASA by freeing up enough resources to stabilize other projects in NASA's core program.[58]

Redesigning for Resilience

Redesigning the Space Station program for resilience is a policy alternative to business as usual. This alternative attempts to improve the return on the public's investment in the Space Station program, and to reduce the probability of termination and restructuring.

The basic idea is to reconceive the annual rephasing exercise as an opportunity to decouple elements of the Space Station program for competitive evaluation. (Decoupling for competitive evaluation reverses the processes of "scaling-up" and putting "all the eggs in one basket.") Those elements that cannot be justified in terms of a particular objective, schedule, and cost would be deleted or deferred. Those elements that survive competitive evaluation would be reconsidered for separate development and operation. They also would be prioritized for separate funding, and developed as warranted by the funds available. It should be possible to stabilize some parts of the program so that they can perform well according to their respective objectives, even if it is not possible to stabilize the program as a whole.

Something close to decoupling has already occurred as an unintended consequence of continuing struggle over the program. NASA has been unable to "build it right" and the opposition has been unable to "not build it at all," but together they have been able to "build it down" to meet annual budget constraints: Block II was decoupled from Block I and deferred indefinitely. Similarly, solar dynamic power, the new space suit, the OMV,[59] the polar-orbiting platform,[60] and other program elements were decoupled from Block I and deleted, deferred, or transferred from the program. However, this "build it down" process is extremely costly, as we have seen, and still has a long way to go before it produces a resilient program.

Redesigning for resilience could begin by anticipating the eventual outcome, the limiting case, of the "build it down" process: The program core that probably would survive in competition with other program objectives and elements, and would be small enough to minimize management and integration problems and to fit within realistic annual budget and total cost constraints. A small, decoupled core could be stabilized through simplification of management and integration under one center, and perhaps through multi-year appropriations if the number of dollars and years are small enough to make such appropriations reasonable. Such a core also might be "built up" efficiently, if not quickly, as program performance and the fiscal situation warrant. In other words, starting from the core, it might be possible to "buy it by the yard" — which is what NASA leaders promised when they promoted the initial concept of the Space Station.

One candidate for the core is a laboratory module, which, in an early 1989 Station baseline, could be man-tended for micro-gravity research before addition of the full 75 kilowatts, a habitat module, and a permanent manned capability. In April, 1989, Chairman Bob Traxler of the House appropriations subcommittee asked about total Station cost "if we stopped the Block I assembly sequence at flight seven or eight — in other words, giving us man-tended capability."[61] According to an insert in the hearing record, the information NASA later provided did not respond to Traxler's request. But Acting Administrator Myers did have an immediate response for Traxler:

> We lose the whole strategy of NASA's future because the manned program is the key to assembly in space, to the operation of laboratory equipment, to the whole program as far as the future is concerned.[62]

A permanent manned capability with the habitat module may be NASA's preference for the core, the last element it would give up under pressure to delete program content.

Later in the same hearings, Traxler pursued the matter with Associate Administrator Odom in a question about the assembly sequence:

> During the man-tended phase of the Space Station - in other words, before the habitat [module] is added to the complex - station has sufficient power to run almost an unlimited number of experiments in the laboratory with an extended duration orbiter. However, once you add the habitat, which comes before you put up the balance of the 37.5 kilowatts of power, the availability to the user drops significantly. Is that correct?[63]

Odom affirmed that was correct. Traxler went on to suggest an automatic hold on the habitat module if a delay were expected in getting the balance of power up, "because it doesn't make a lot of sense to add infrastructure and take away power from the user." Odom replied, "That is certainly an option, but if you want to get permanently manned capability as soon as possible, then obviously, the habitat is mandatory."[64]

The FY 1991 appropriations bill recently passed by the House of Representatives appears to protect the power supply for users of a man-tended U.S. laboratory module. It stipulates that at least 30 kilowatts must be available to users of the Station, and that full power of 75 kilowatts must be provided before any international modules are added to the Station.[65]

Elimination of the Space Station truss would facilitate if not force decoupling of the modules for separate evaluation. This is not out of the question. A NASA task team recently recommended that the agency consider "the total elimination of the truss and its replacement with pressurized modules in which, and on which, all station elements would be mounted..."[66] This was proposed as a means of reducing the EVA time required for external maintenance. Elimination of the truss would also focus attention on finding the core.

Redesigning for resilience might also include smaller space stations as an alternative to finding the core of the present Space Station. Representative Bill Green, the ranking minority member of the House appropriations subcommittee, contended that "The time has come for us to consider scrapping NASA's plan and replacing it with an updated version of our former space station – Skylab."[67] Green went on to mention another alternative without recommending it:

> Several years ago, we nearly approved a smaller version of Skylab f many of you will remember the Industrial Space Facility - the ISF. This program, which I personally thought to be a wonderful idea - and a test bed for the Station, was scrapped when it became perceived as a threat to the Space Station - another victim of political shortsightedness.[68]

Redesigning for resilience entails short-run instability as the program is taken apart in order to achieve stability later on a smaller scale. For Space Station supporters who believe that the program can be stabilized before termination, redesigning for resilience is an inferior alternative to business as usual. For supporters who believe that business as usual means eventual termination and perhaps restructuring, redesigning for resilience may merit consideration.

However, NASA controls the technical expertise necessary to design and assess in sufficient detail the alternatives to business as usual. OMB and Congress are effectively limited to requesting alternatives and responding to whatever NASA provides. (The ISF is a partial exception.) Consequently, NASA is in the driver's seat: NASA can stonewall or avoid responding to requests for alternatives to business as usual, leaving OMB and Congress with a choice between the present Space Station as it is built down incrementally, or no space station at all.

Summary and Conclusion

The scenarios and policy alternatives considered here provide a frame of reference for thinking through the future of the Space Station program as events unfold. The key is to look beyond claims that the program will soon be stabilized to the evidence and logic that bears on those claims. The rationality and political feasibility of forthcoming decisions on the program depend most critically on the question of stabilization.

We contend that the present Space Station program cannot be stabilized over the next several years, despite the best efforts of NASA, the Administration, and Congress. The underlying problem is that the program lacks the resilience necessary to perform as promised in a turbulent situation that NASA cannot anticipate or control. The problem can be traced to the 1984 concept of the program — large-scale, long-term, and tightly coupled with other programs — which subordinated program performance to other considerations.

If so, there are no good alternatives for resolving the Space Station issue. Business as usual, the attempt to stabilize the program and to cultivate the appearance of program stability, leads either to a risky space station — less capable, later, and more costly than the present one — or to program termination and possibly the restructuring of NASA. Redesigning for resilience could salvage something more quickly and at less cost from the present program, or replace it with a smaller space station program. But all those outcomes that appear to be feasible are also failures with respect to the capabilities, schedules, and costs that were promised in 1984 and are promised in the current baseline.

The larger issue is whether, or under what circumstances, NASA will recognize the need for resilience, and design it into future programs and into civilian space policy as a whole. The alternative is to replicate the mistakes of the Space Station program throughout the civilian space program. This is a particularly disturbing possibility while the Earth Observing System enters development and the unformed Space Exploration Initiative continues to incubate.

Notes

1. The authors gratefully acknowledge the comments of many authors of chapters in this volume, particularly the comments of Albert D. Wheelon, and the support of the Alfred P. Sloan Foundation.

2. Bob Davis, "After Recent Setbacks, NASA Faces a Fight Over Project Funding," *Wall Street Journal* (July 9, 1990), p. A1.

3. *Space Station News* (August 20, 1990), p. 4.

4. An earlier paper attempted to define the underlying problem in the program as a necessary first step toward understanding how the controversy might be resolved. See Ronald D. Brunner and Radford Byerly Jr., "The Space Station Programme: Defining the Problem," *Space Policy* 6 (May 1990), pp. 131-145.

5. A secondary theme in the testimony is that content deleted from budget requests -- such as Block II or solar dynamic power -- has been deferred, not permanently deleted. Thus program capabilities are alleged to be retained.

6. Of course stabilization allows for normal changes and refinements within the envelop defined by fixed design capabilities, schedule, cost.

7. *1989 NASA Authorization*, Hearings before the Subcommittee on Space Science and Applications, U.S. House of Representatives (March 10, 1988), pp. 220, 221.

8. See NASA, Office of Space Station, *Space Station Capital Development Plan Fiscal Year 1989* (April, 1988), which was submitted to the House Committee on Science, Space and Technology and the Senate Committee on Commerce, Science and Transportation.

9. *1990 NASA Authorization*, Hearings before the Subcommittee on Space Science and Applications, U.S. House of Representatives (February 28, 1989), p. 288.

10. See his testimony in *Proposed Space Station Freedom Program Revisions*, Hearings before the Committee on Science, Space, and Technology, U.S. House of Representatives (October 31, 1989), p. 20. See also NASA's response to a question submitted in advance of the hearings, p. 78: "In the course of our review, some station subsystems, the technical and schedule risk were found to be sufficiently high to merit a reassessment even if the program had received its full funding."

11. *Ibid.*, p. 12.

12. Statement by Dr. William B. Lenoir prepared for the Subcommittee on Space Science and Applications, House of Representatives (February 21, 1990), pp. 22-23.

13. Chairman Roe of the House authorization committee characterized the rephased Station as "a radical departure from the design and schedule that this committee has given its unwavering support to." See *Proposed Space Station Freedom Program Revisions, op. cit.*, p. 3.

14. "Station Schedule Slips, Cuts Likely," *Space Station News* (June 25, 1990), p. 3.

15. Sometimes NASA spokesmen imply, erroneously, that full funding or funding stability is a *sufficient* condition for program stability. See, for example, prepared testimony in *Proposed Space Station Freedom Program Revisions, op. cit.*, p. 93: "It is critical that some stability be introduced into Space Station funding in order to avoid program disruptions and future deferrals or deletions."

16. See notes 9 and 13 above and *Proposed Space Station Freedom Program Revisions, op. cit.*, p. 78, for the most relevant testimony. These specific changes included switches to DC power distribution and hydrazine propulsion modules, and deferral or deletion of the new space suit and the closed loop environmental control life support system.

17. *Proposed Space Station Freedom Program Revisions, op. cit.*, p. 23.

18. *Ibid.*, p. 2.

19. Roe as quoted in *Defense Daily* (July 6, 1990), p. 30.

20. *Proposed Space Station Freedom Program Revisions, op. cit.*, p. 4.

21. *Ibid.*, p. 7.

22. Selective inattention is not unusual. In non-technical terms, selective inattention means "you just miss all sorts of things which would cause you embarrassment...to notice. It is the means by which you stay as you are, in spite of the efforts of worthy psychiatrists, clergyman, and others to help you mend your ways. You don't hear, you don't see, you don't feel, you don't observe, you don't think, you don't this, you don't that, all by the very suave manipulation of the contents of consciousness by anxiety..." From Harry Stack Sullivan, *The Fusion of Psychiatry and Social Science* (New York: W. W. Norton, 1964), pp. 216-217.

23. For more on the lack of resilience as the underlying problem in the Space Station program, see Brunner and Byerly, *op. cit.*

24. David E. Rosenbaum, "Estimate on Deficit Reduction Is Raised Sharply," *New York Times* (July 17, 1990), p. A8.

25. David A. Stockman, *The Triumph of Politics* (New York: Harper & Row, 1986); and Alice M. Rivlin, "A Public Policy Paradox," *Journal of Policy Analysis & Management* 4 (1984), pp. 17-22.

26. *Congressional Record* (June 28, 1990), p. H4360.

27. These include, in addition to requests for multi-year appropriations, transfers of some Space Station program costs to other NASA programs and to other departments (particularly DOD), and attempts to develop some program elements with commercial funds, as well as attempts to lobby and mobilize support for higher appropriations through letters from aerospace contractors, the circulation of figures on the amount and geographical distribution of program funds and employment, and even TV specials.

28. According to a recent report by Bob Davis, *op. cit.*, p. A6, "the operation of the station itself will require so much power that so far it has only 10 to 15 kilowatts available for experiments -- less than the 30 considered adequate for the labs."

29. See William J. Broad, "Space Station Must Be Altered To Reduce Work, NASA Says," *New York Times* (July 21, 1990), p. A1. For background see William J. Broad, "Major Flaw Found in Space Station Planned by NASA,' *New York Times* (March 19, 1990), p. A1; William J. Broad, "Costs Stir Doubt If Space Station Can Be Kept Up," *New York Times* (July 11, 1990), p. A1; Andrew Lawler, "Team Finds Too Many Space Walks Needed to Tend Station," *Space News* (July 9-15, 1990), p. 3.

30. See "Station Weight Becomes Number One Concern," *Space Station News* (July 9, 1990), pp. 1-5. See also *Defense Daily* (July 10, 1990), pp. 42-43 and (July 11, 1990), p. 54.

31. This volumetric problem is discussed in connection with the weight problem, for which additional Shuttles are a possible solution. See the sources cited in the previous footnote. See also, the National Research Council, *Space Station Engineering Design Issues* (Washington, D.C.: National Academy Press, 1989), in which a committee reports, p. 49, that "the 50-ft photovoltaic radiator will not fit in the Shuttle payload bay for the first flight."

32. U.S. General Accounting Office, *Space Debris a Potential Threat to Space Station and Shuttle*, GAO/IMTEC-90-18 (April 1990).

33. Leonard Davis, "Space Station Power Plan Raises New Design Concerns," *Space News* (July 23-29, 1990), p. 17.

34. H. Buecker and R. Facius, "Radiation Problems in Manned Spaceflight with a View Towards the Space Station," *Acta Astronautica* 17 (1988), pp. 243-248.

35. A bridge crane accidentally flexed Discovery's right payload door. *Aviation Week & Space Technology* (July 9, 1990), p. 11.

36. William J. Broad, "U.S. Space Agency Grounds Shuttle Over a Fuel Leak," *New York Times* (June 30, 1990), p. A1.

37. Office of Technology Assessment, *Round Trip to Orbit: Human Spaceflight Alternatives -- Special Report* (August 1989), p. 6.

38. This is a conclusion of the Space Station Advisory Committee, as quoted in *Space Station News* (June 11, 1990), p. 8. Later that month, the optical problem of the Hubble Space Telescope was traced to the failure to test the mirrors in combination on the ground. See Warren E. Leary, "Hubble Telescope Loses Large Part of Optical Ability," *New York Times* (June 28, 1990), p. A1.

39. Leonard Davis, *op. cit.*

40. The GAO report on orbital debris, *op. cit.*, p. 18, which went on to report that "A senior NASA engineer voiced his opinion that NASA is wasting time, money, and resources by developing the station to meet understated design requirements."

41. Leonard Davis, *op. cit.*

42. "Station Weight Becomes Number One Concern," *Space Station News* (July 9, 1990), p. 4.

43. In an interview in *Space Station News* (August 20, 1990), p. 7, former NASA Administrator James Beggs acknowledged it was a mistake not to put a single center in charge of the Space Station program. That was done for what he described as "political reasons."

44. *Departments of Veterans Affairs and Housing and Urban Development, and Independent Agencies Appropriations for 1991 -- Part 4, National Aeronautics and Space Administration*, Hearings before a Subcommittee of the Committee on Appropriations, U.S. House of Representatives (March 20, 1989), p. 80.

45. "NASA Authorization," Hearings before the Subcommittee on Science, Technology, and Space, U.S. Senate (March 16, 1989), p. 96.

46. For an example, see Bob Davis, *op. cit.*

47. The seminal statement of the vision is Wernher von Braun, "Crossing the Last Frontier," *Collier's* (March 22, 1952), pp. 24f.

48. Senator Gore, Hearings before his Subcommittee on Science, Technology, and Space (March 16, 1989), p. 1. To the witnesses, he said (p. 2) that "each of you must work harder to make the case for the space station program as a budget priority, as a foreign policy priority, and as a technological leadership priority."

49. Representative Nelson, *1990 NASA Authorization*, Hearings before the Subcommittee on Space Science and Applications (February 28, 1989), p. 277.

50. Representative Charles Schumer introduced amendments to transfer funds from the Space Station program in 1988 and 1989. Floor action on the latter occasion, the twentieth anniversary of the first moon landing, can be found in the *Congressional Record* (July 20, 1989), pp. H3960-H3961, H3979-H3993. Floor action on a similar amendment introduced by Senator Daniel Patrick Moynihan can be found in the *Congressional Record* (July 12, 1988), pp. S9366-S9377. More recently, in a July 25, 1990 address to the National Press Club, Representative Bill Green advocated termination of the Space Station in favor of an advanced Skylab.

51. For some precedents and theory, see H. W. Lambright and H. M. Sapolsky, "Terminating Federal Research and Development Programs," *Policy Sciences* 7 (1976), pp. 199-213.

52. Notice, for example, how the conjunction of the Hubble's optical problem, the Shuttle's hydrogen leak, and the Station's EVA and weight problem was generalized into concern about deeper problems in NASA. In that frame of reference, even the delay of an Atlas launch is network news.

53. Hearings before the House appropriations subcommittee (April 25, 1989), p. 38.

54. *Space Station News* (July 17, 1989), p. 4.

55. See the *Report to the President* by the Presidential Commission on the Space Shuttle Challenger Accident (June 6, 1986), p. 200. Richard P. Feynman, a member of the Commission, concluded that "because of the exaggeration at the top being inconsistent with the reality at the bottom, communication got slowed up and ultimately jammed." From *What Do You Care What Other People Think?* (New York: W. W. Norton, 1989), p. 215.

56. Until recently, the Savings & Loan crisis illustrates the potential of the Administration, the Congress, and the public to overlook promises not fulfilled and problems not resolved.

57. Hearings before a Subcommittee of the House Appropriations Committee (April 25, 1989), p. 72.

58. See David H. Moore, "A Budget-Constrained NASA Program for the 1990s," in Radford Byerly, Jr., ed., Space Policy Reconsidered, ch.1 (Boulder: Westview Press, 1989).

59. See "OMV Program Gets the Ax," *Space Station News* (June 11, 1990), p. 1.

60. See "Major Changes Ahead for Polar Platform, WP 3," *Space Station News* (January 22, 1990), p. 1.

61. Hearings before a Subcommittee of the House Appropriations Committee (April 25, 1989), p. 46.

62. *Ibid.*, p. 47.

63. *Ibid.*, p. 59.

64. *Ibid.*

65. As reported in *Space Station News* (July 9, 1990), pp. 6-7.

66. William F. Fisher and Charles R. Price, *Space Station Freedom External Maintenance Task Team, Final Report* (July, 1990), Vol. I, Part 1, p. 78.

67. From remarks prepared for an address to the National Press Club in Washington, D.C. on July 25, 1990, p. 5.

68. *Ibid.*, p. 6.

Chapter 14
THE SPACE SHUTTLE PROGRAM: PERFORMANCE VERSUS PROMISE

Roger A. Pielke, Jr., and Radford Byerly, Jr.

THE WHOLE AIM OF THE SCIENTIFIC STUDENT OF SOCIETY IS TO MAKE THE OBVIOUS UNESCAPABLE... THAT WHICH IS KNOWN IMPLICITLY AND BASED UPON DIFFUSED, UNVERBALIZED EXPERIENCE MUST BE MADE EXPLICIT IF NEW WAYS OF DEALING WITH THE WORLD ARE TO BE INVENTED.
—H. D. LASSWELL

NEVER PROMISE MORE THAN YOU CAN PERFORM.
—LIVY

Introduction

On February 2, 1989 James Fletcher, then NASA Administrator, testified before a Congressional committee that "the [Space] Shuttle is the fundamental link in a space transportation infrastructure which will carry this nation into the 21st century. It is our means to place human beings into orbit and it makes Space Station Freedom possible and practical".[1] More recently, in an official NASA publication a similar view of the Shuttle program was given: "Any spacefaring nation would characterize an optimum transportation system as being a reliable, reusable, man-rated, heavy-lift booster, carrying crews and significant cargos to and from space – the very definition of the Space Shuttle program".[2] NASA officials have regularly characterized the Shuttle as a "fundamental link" in the United States civilian space program.[3] However, a well-respected committee of experts, who were asked to examine NASA, recently suggested in a highly publicized report that the United States should "defer or eliminate the planned purchase of another orbiter" based on the committee's concern that "the space shuttle would seem to be the weak link of the civil space program -- unpleasant to recognize, involving all the uncertainties of statistics, and difficult to resolve".[4] This viewpoint is very different from that espoused by NASA.

While it is not logically inconsistent for a "weak link" to also be a "fundamental link", certainly it would defy logic to make a weak link fundamental to our space program. This paper provides performance information on the Shuttle relevant to whether or not it is a weak link. The information can also illuminate decisions on other programs -- Space Station, Human Exploration, a new launch system -- which depend on evaluation of Shuttle performance. Better information is the first step toward better policy.

Cost, schedule, and capability are the factors for analysis because upon project approval these factors are established as programmatic goals, thus providing a baseline against which performance may be measured. Perhaps more importantly, the Shuttle program's initial promises were implicitly agreed between NASA and Congress.[5] As will be shown, the difference between initial promise and actual performance suggests that the Shuttle program as originally conceived by NASA and approved by Congress was poor public policy.[6]

As a necessary first step towards better policy this paper documents the performance shortfall. To fully understand the policy shortfall it would be necessary to conduct a full policy appraisal of the Shuttle program; for example expanding the analysis to include examining *why* the gap between promise and performance exists, assessing likely future scenarios for the Shuttle program, and recommending the most worthwhile policy alternative for the program. A full policy appraisal goes beyond the scope of this paper.

It is hoped that performance that falls well short of program promise is perceived to be a significant policy problem, stimulating future work to examine why this occurs and how we may do better in the future -- in the Space Shuttle program as well as in other space programs.

The factors for analysis are defined as follows:

Cost
The *total cost* of the program, as measured from program inception to termination.[7]

The *average cost* of a Shuttle flight, as measured in several different contexts. A range of future projections of the program's average cost per flight, as well as historical data, will be presented.[8] Furthermore, costs associated with Shuttle *attrition rate* will be considered.

The *annual cost* of the program is relevant as funding decisions are made on a year-by-year basis. Here as well, both historical data and future projections will be considered.[9]

Schedule
An analysis of the *flight rate* is necessary to ascertain what may be expected from, as opposed to planned for, the program in future years.

Capability
The *capability* of the Space Shuttle may be evaluated in many ways, both quantifiable and nonquantifiable, with respect to programmatic goals. Measures of capability include: payload capacity, payload type (e.g., man), etc. Measurements of performance are necessary as decisions are made concerning the future of the civil space program.

The twenty year history of the program and 10 years of Shuttle operations provide data for an examination of the program based on these factors. Because the data are difficult to collect and interpret, the methods of analysis used are purposely simple and transparent. These methods are no substitute for more sophisticated methods, e.g., economic and engineering, but rather are preliminary and complementary. Better data will support more sophisticated analysis of this and other civil space programs.
    This paper builds on studies of Shuttle history by examining *how* program performance compares to original goals, and what might be expected for the future.[10] Rather than dismissing program shortfalls we should strive to understand them.
avoid making the same mistakes again. In 1972 NASA predicted the net cost of the Space Shuttle program as shown in Table 14.1.[11]
    One must add $8 billion in 1990 dollars to the $42.7 billion NASA estimate to represent civil service salaries which are not accounted for in the table.[12] Thus the total cost estimate for the Shuttle program is $50.7 billion through 1990.

Table 14.1. NASA predictions for the total cost of the Shuttle program through 1990 (dollars in billions).

	Then-year dollars	1990 dollars
1. Investment in Space Shuttle, including initial inventory (details show below)	$6.45	$17.1
(a) Develop, test, and procure 2 orbiters and 2 boosters (1972-1980)	*(5.15)*	*(13.6)*
(b) Refurbish 2 orbiters and procure 3 more, including engines, and initial production boosters (1979-1980)	*(1.0)*	*(2.6)*
(c) Facilities for development, test, launch and landing capability	*(.3)*	*(.8)*
2. Additional investments required to fly mission modes assumed; (includes possible future development of space tug by 1985, expendable injection stages for high orbit satellites and deep space probes prior to tug development, and operational site facilities at Vandenberg AFB, California.)	$1.6	$4.2
3. Total launch costs, including procurement of replacement boosters, 580 flights (1972-1990)	$8.1	$21.4
4. Total 1972-1990 (sum items 1, 2, and 3)	$16.15	$42.7

NASA also made predictions for the total number of flights that were to occur during the same time period. From 1979-1990 the Shuttle was to fly 580 flights, about 48 per year. NASA's planning model suggested that the 580 Shuttle flights would be so much cheaper than 580 equivalent conventional launches (i.e., on expendable launch vehicles or ELVs) that the Shuttle would pay for its development, its operations, and still save over $13 billion ($5 billion in 1972 dollars).

1972: Shuttle Program Promises

Perhaps because there is no dispute over whether the Space Shuttle program achieved the goals set out in 1972 there is not much discussion over how the program fails to meet them. Nevertheless, it is important to remember those goals: Not to lay blame, but rather to illuminate how these goals were not met and to develop strategies to Later there were plans to recoup launch costs by charging for the use of the Shuttle.[13] Other performance goals were mainly payload-based. In order to fill almost 50 payloads per year NASA listed a range of possibilities from getaway specials, i.e., small payloads flown on a space-available basis, to large scale space construction. The goals for the project set out in 1972 are clear and unambiguous, thus providing criteria for program appraisal.

Shuttle Program Costs: 1970-1990

In any public enterprise project cost is important to policymakers, policy analysts, and American citizens as projects often compete on that basis. Project cost, both historical and projected, can provide some of the information necessary for both program appraisal and policy decisions.

The manner which NASA keeps its accounts does not facilitate this type of analysis because costs are not accounted by specific programs, but rather by budget categories. Shuttle costs have been identified in the following four budget categories:[14]

Research and Development [R & D] : This category primarily includes the costs budgeted for the development phase of the project under the line item *Manned Space Flight: Space Shuttle* from fiscal year (FY) 1972 to FY 1984. After 1984 there are additional costs in this category, e.g., in *Engineering and Technical Base, Payload Operations and Support Equipment*, and *Spacelab*, but we have not included them because they are difficult to separate and identify. Expendable launch vehicle costs, budgeted in this category, have been excluded.

Space Flight, Control and Data Communications [SFCDC] : This category primarily includes Shuttle costs budgeted for the operational phase of the project in the following two line items: *Shuttle Production and Operational Capability* and *Space Transportation Operations* from FY 1985 to FY 1990.[15] There are also tracking and communications costs in this category, which were not included because we know of no way to allocate them to Shuttle.

Construction of Facilities [CoF] : This category includes all costs which are explicitly labeled in the NASA budget for the Shuttle program for the purpose of providing for repair, modification, new construction, and design and planning of facilities.

Research and Program Management [R&PM]: This category includes costs for civil service staff, i.e., salaries, maintenance of facilities, and technical and administrative support. Unfortunately, within this budgetary line item costs are not broken down by program, but rather by NASA center. To arrive at the R&PM costs for the Shuttle program we made use of the available data as follows: Shuttle R&PM costs were taken to be equal to the total R&PM costs at a NASA center multiplied by the fraction of that center's personnel that work on the Shuttle.[16] This calculation was done for the three main centers where Shuttle work is done: Kennedy Space Center, Marshall Space Flight Center, and Johnson Space Center. Additional Shuttle R&PM costs at other centers were not included.

As noted some costs were not included due either to the difficulty of accurately ascertaining the Shuttle-relevant part, or their insignificance. The addition of these categories would increase the estimates, but probably by less than 10%. Also not included is the cost-of-money. Thus, again the estimates are low.

The Department of Defense [DoD] also spent a significant amount of money on the Space Shuttle program which is not included. Through 1980 DoD had spent $1.8 billion on the Shuttle program, and in 1980 it was estimated that an additional $1.2 billion would be spent by 1984. The funding was to cover the development of the Inertial Upper Stage [IUS], construction of the Shuttle launch complex at Vandenberg Air Force Base, DoD launches, and all other DoD Shuttle-related costs.[17]

NASA also receives funds for the Shuttle program when it is reimbursed for a Shuttle flight, e.g., when it flies a DoD, commercial, or foreign payload. These reimbursements represent additional resources available to the program -- they do not go to the Treasury to offset the costs of the program. These funds were also not included in the tabulation. In FY 1990 the Shuttle reimbursables were $82 million, compared to slightly more than $4 billion which were appropriated.[18] In FY 1985 when there was much more reimbursable activity (e.g., DoD and commercial payloads) NASA was reimbursed about $480 million for the Shuttle, compared to appropriations of almost $4 billion.[19]

Total Costs. Figure 14.1 shows that the total cost of the Space Shuttle program through 1990 is approximately $65 billion, compared to the original estimate of $51 billion. While this is a fairly small overrun, about 27%, it must be remembered that the $51 billion was to have paid for 580 flights while only 37 had been flown through 1990. The R & D and SFCDC categories make up most of this amount, about 83%. The Construction of Facilities line item for the Shuttle program is relatively small, on the order of 2% of the total. Research and Program Management accounts for approximately 15% of the total.

Annual Costs. As funding decisions are made on a year-by-year basis in Washington the program's annual cost is important to consider. Figure 14.2 displays this data. Since 1982 the Shuttle program's annual costs have been approximately $4 billion, although costs seem to be rising.[20] Of note here is that annual costs during the "operational" phase, i.e., beginning with the fifth flight, have been higher than during the pre-flight "development" phase. Costs were highest in FY 1982 and FY 1983 when there were still significant orbiter construction costs as well as large flight costs. Budget estimates for the next three years show annual program costs rising to about $5 billion in 1993.

Average-cost-per-flight. The subject of much debate over the lifetime of the Shuttle program has been how much a Shuttle flight actually costs. No attempt is made here to ascertain how much the *next* flight costs, i.e., the marginal cost, but rather to determine how much the average cost of a Shuttle flight will have been over the entire life of the program based on various estimates of when it will be terminated.

If the program had ended with the completion of FY 1990 the average cost per flight would have been about $1.7 billion. Ignoring the costs incurred prior to the "operational" phase - i.e., before 1983 - the average-cost-per-flight would be lowered to about $1.1 billion. For comparison, Macauley and Toman calculate a long run marginal cost ranging from $200-320 million (1990 $) per flight, and the GAO has recently used an "additive" cost of approximately $80 million.[21]

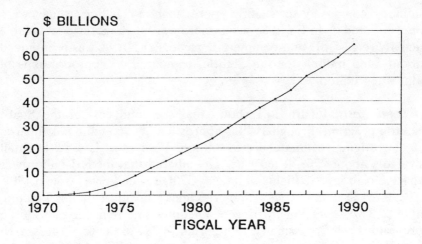

Figure 14.1: Space Shuttle Program Total Costs. Cumulative cost of the Shuttle program in 1990 dollars. Note: Transition quarter not included. Source: NASA.

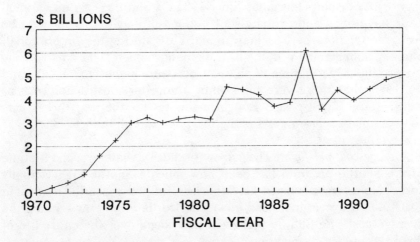

Figure 14.2: Space Shuttle Program Annual Costs. Annual costs of the Shuttle program in 1990 dollars. Note that costs are now higher than during the development phase, 1972-1981. High costs occurred in FY 1982 and 1983 when there were both significant operations costs and orbiter construction costs. the 1987 peak is an anomaly due the one-time replacement for the *Challenger*. Note: FY 1991-93 estimates. Source: NASA.

Figures 14.3a and b show the average cost per flight over the entire history of the program for a range of possible futures.[22] Two program termination years are considered: 2000 and 2010. For each of these scenarios the average cost per flight is calculated at the end of the relevant time period for a range of 2 to 12 flights per year, assuming average annual future appropriations of $2, $4, or $6 billion per year to operate the program.

These figures show that for the reasonable projection of 8 flights per year and a $4 billion average annual appropriation, the average cost per Shuttle flight will be about $800 million for either termination year. This corresponds to an average cost per flight of $500 million for the remainder of the Shuttle program, independent of termination date.[23] This is an important cost number for making decisions about the program, as it does not include sunk costs. More pessimistic projections of annual appropriations and flight rates push the average cost per flight over $1 billion.

Although in principle the average cost of a flight is important for program evaluation, it has much less political importance. Politically the Shuttle is a highly visible, manned space program, a symbol of U.S. leadership. Only secondarily is it a transportation program. Thus, its costs are seen as costs of having the program, not costs of getting payloads to orbit. Thus, it makes sense for policymakers to think "if we have the Shuttle we may well fly it as often as possible". Therefore, from this perspective the Shuttle may be considered "free" transportation, at least for NASA payloads. Furthermore, if the shuttle is (or will be) flying as often as possible, then it makes little sense to speak of marginal costs as there is no margin. However, the marginal cost of one less flight becomes important if the program is to be considered operating at less than maximum capacity.

Not included in the average cost calculations presented above is the cost of replacement orbiters. The Challenger replacement, Endeavor, cost $2.3 billion which was appropriated in a lump sum.[24] If lost orbiters are to be replaced in the future, then the average cost per flight of orbiter attrition can be determined by assessing a predetermined "surcharge" on each flight based on the expected reliability. The Office of Technology Assessment has estimated the Shuttle's reliability to be 0.98. If this is accurate there would be a 50-50 chance of losing at least one orbiter within 34 flights, or in other words, within 3 to 6 years depending on the flight rate.[25] Based on

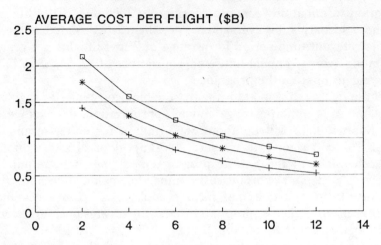

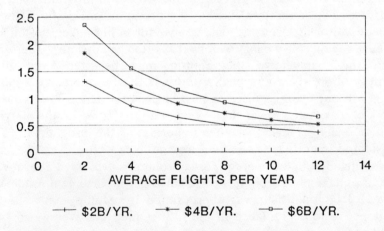

Figures 14.3a and b: Average Cost per Shuttle Flight at Program Completion. The average cost per Shuttle flight calculated assuming various numbers of flights per year and assuming three different average levels of funding per year, $2 B, $4 B, and $6 B. Figure 14.3a assumes the program will end in the year 2000, while 3b assumes it will end in 2010, i.e., that there will be more flights over which to spread early costs.

this estimated reliability, and assuming lost orbiters will be replaced for $2.3 billion, one estimate of the surcharge per flight to pay for orbiterattrition is $46 million.[26] Since orbiter reliability is not 1.0, policy should reflect the likelihood of losing orbiters.

Shuttle Schedule: 1981-1990

Schedule refers to when capabilities are available. The first manned orbital flight of the Shuttle was originally planned for 1978 and actually occurred in 1981. Flights were planned to occur regularly after that, at an average rate of nearly 50 per year.

For two reasons the best data to analyze the Space Shuttle flight rate are from the pre-Challenger era. First, 25 successive launches occurred forming the longest unbroken stretch of launches in the Shuttle's history. Secondly, since the Challenger accident the Shuttle program has performed less well due to a cautious restart period and numerous down-times, and therefore is not representative of what the Shuttle has done. The pre-Challenger data can now be considered a best-case scenario as NASA no longer intends to pursue such a high flight rate. It should be emphasized that the data presented here are not for planning purposes, but instead are a reasonable estimate as to what policymakers might expect in the future.

Analysis of the Shuttle flight rate is made difficult due to the number of variables involved. Factors which have an impact on the flight rate include:[27]

NASA Budget : The annual appropriation from Congress, which determines overall resources.

Mission Payloads : The payloads carried to orbit, e.g., the Hubble Space Telescope. Some payloads take more time to integrate, some have problems that cause delays.

Vehicle Assembly Building (VAB) : Here the orbiter is mated to the solid rockets and the external tank. Orbiters must pass through the VAB serially, thus, unexpected problems can cause delays on subsequent flights.

Number of Orbiter Processing Facilities [OPF] : Much of the turn-around work on the orbiters is done in the OPFs. When there are fewer OPFs than orbiters, a problem with one orbiter could stall another.

Number of Mobile Launch Platforms [MLP] : These ferry the Shuttle from the VAB to the launch pad.

Landing Site : Orbiters which land at Kennedy can be prepared for launch in less time than ones which land at Edwards Air force Base, due to the time it takes for transfer from Edwards to Kennedy.

Shuttle Transfer Aircraft : Modified 747s that return orbiters from landing sites to Kennedy Space Center.

Mission Duration : This is the time from launch to landing.

Weather : This can be either at Kennedy or transatlantic abort landing sites, potentially delaying launch, or at Edwards, White Sands, or Kennedy, affecting landing.

Payload Launch Window : This is time period within which a payload must be launched in order to meet mission objectives, e.g., planetary alignments.

Crew Training Time : This is the preparation time necessary for the Shuttle crew. It cannot proceed ahead of development of mission software, which in turn depends on payloads, etc.

Logistics : Primarily the availability of spare parts to repair Shuttles.

Hardware Failures : Including wear-out due to cumulative use, as well as accidents.

Ground Crew Time-off : The time periods during which no work is done on Shuttles due to holidays, etc..

Which Orbiters Fly : All orbiters are not equally capable of sustaining the same flight rate. For example, it has been generally recognized that Columbia does not perform as well as the other orbiters.

Shuttle Hardware : These include the orbiter, the Solid Rocket Boosters, the External Tank, the Main Engines, and other hardware necessary for flight.

It is necessary to distinguish between the number of orbiters and the number of "schedulable" orbiters. Because orbiters occasionally have to be removed from their flight rotation for major repairs or factory modifications, the number of schedulable orbiters is less than the actual number of orbiters. In 1986 the National Research Council found that of the three orbiters there existed only "a bit in excess of 2" schedulable orbiters, which gives an estimated schedulability ratio of about 0.7 per orbiter.[28]

Another necessary distinction is between work days and calendar days: Work days are defined as those in which normal orbiter preparation takes place. For example, during the one-year period

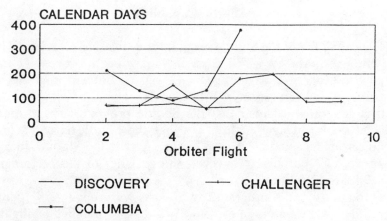

Figure 14.4 Space Shuttle Program Orbiter Turn-Around Time. Orbiter turn-around time (i.e., launch to launch) in calendar days, for flights before the *Challenger* accident.

beginning with the Shuttle flight of August 30, 1984 NASA launched 9 missions using only two orbiters, Challenger and Discovery. The two orbiters were able to achieve an average turn-around of 55 work days. However, this amounted to an average turn-around time of 81 calendar days between flights. The difference of 26 days is accounted for by ground crew time-off, mission duration, transfer to Kennedy from the landing site, weather delays, on-the-pad payload problems, and other such factors and contingencies.

Figure 14.4 shows some data on pre-Challenger turnaround experience for the three orbiters which flew most of the flights.[29] Ground operations for the first flight of each orbiter are ignored.[30] Worth noting is the poor performance record of Columbia, the oldest active orbiter. As long as Columbia is in the fleet overall flight rates may be less than would be possible without it.

After the Challenger accident NASA's goal has been to achieve a turnaround time of 75 work days per flight.[31] Adding 25 days to represent other time factors results in a best-case figure of 100 calendar day orbiter turnaround. Therefore, the post-Challenger best case scenario, assuming 4 orbiters (2.8 schedulable) is 10.2 flights per year. If the fraction of schedulable orbiters were raised to 0.8 (e.g., by decreasing time for factory modifications or avoiding accidents) the annual flight rate could be 11.7 per year, consistent with a 1988 OTA estimate of 12 to 14 flights per year.[32]

Shuttle Capability

The capability of the Shuttle program is difficult to measure. There are three criteria by which capability may be measured: (1) with respect to program's original goals, (2) against similar programs, and (3) with respect to present or future goals.

It is generally accepted that the Shuttle program has not met its original goals. This is not to say that the Shuttle is not a grand technical achievement -- it is. Nor to say that the Shuttle does not serve a useful purpose. However, the specifics of the performance shortfall are as follows.

Through 1990 the program's total costs were about 27% percent more than what was predicted in 1972.[33] The number of flights is one fifteenth the 1972 prediction. The average cost per flight is over 19 times what was promised. It is worth noting that if the Shuttle were to fly an average of 8 flights per year, at $4 billion average annual appropriation, it would take about 68 additional years, and approximately $270 billion, to reach the original goal of 580 flights. In addition, the Shuttle recovered few of its operating costs during the 1980s, and will not recover many in the future due to a policy decision against flying commercial flights and DoD's decision to minimize its Shuttle flights.

Figure 14.5 shows actual Shuttle performance (solid line, with circles showing individual flights) versus various predicted flight plans (dotted lines).[34] It shows that predictions were repeatedly missed. Each delay -- the horizontal distance between prediction and actuality -- represents unaccounted costs and disappointments to payload operators.[35] Data on the costs of delays are not readily available. It now seems that a more achievable flight rate is planned. However, the next major task for the Shuttle is assembling the Space Station, which will require some twenty flights be flown in sequence as scheduled years in advance, which has never occurred.

Typically Shuttle flights are flown in an order and on dates very different from those originally planned. For example, reading from the January, 1984 manifest, the fifteen flights 10 through 24 were scheduled to fly in that order, i.e., 10, 11, 12...24. The next fifteen flights actually flew in the following order: 10, 11, 12, 14, 16, 18, 19, 17, 22, 21, 24, 25, 27, 29, and 30. Table 14.2 shows how the launch dates planned in various published Shuttle manifests for representative payloads become disordered. Mostly the launch dates slip (delay) but occasionally they move ahead, or move to a different orbiter. This is

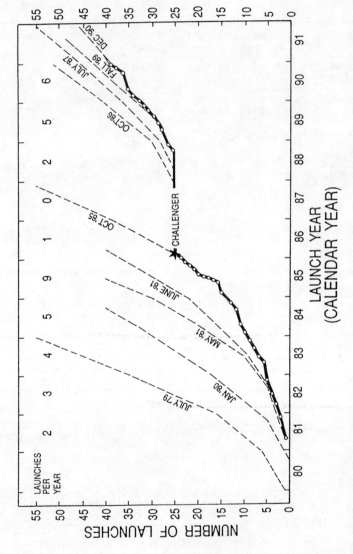

Figure 14.5: Shuttle Schedule vs. Actual Performance. Shuttle manifest plans versus actual launches. Flight delays are shown by the horizontal distance between manifest (dotted line) and actual (solid) for a given flight number. Note that more manifests were planned than could be plotted on this graph.

Table 14.2. Launch date and vehicle for 11 primary payloads as scheduled on 6 different manifest dates. n.a. = Not Available.

LAUNCH SCHEDULE FOR VARIOUS SHUTTLE PAYLOADS
(Payload Manifest Date)

Payload	8/84	6/85	10/86	8/88	1/90	12/90
LDEF RET	3/18/85 Discovery	9/24/86 Challenger	11/15/90 Atlantis	11/13/89 Columbia	1/9/90 Columbia	Launch 1/9/90
HST	8/6/86 Atlantis	8/8/86 Atlantis	11/17/88 Atlantis	2/1/90 Atlantis	4/18/90 Discovery	Launch 4/24/90
ASTRO-01	3/6/86 Columbia	3/6/86 Columbia	1/19/89 Columbia	3/1/90 Columbia	5/9/90 Columbia	Launch 12/12/90
SLS-01	12/22/86 Challenger	n.a. n.a.	12/7/90 Discovery	6/7/90 Columbia	8/29/90 Columbia	5/91 Columbia
ULYSSES	5/15/86 Challenger	5/15/86 Challenger	n.a. n.a.	10/5/90 Atlantis	10/5/90 Discovery	Launch 10/6/90
GRO	5/88 Atlantis	n.a. n.a.	1/18/90 Columbia	4/5/90 Discovery	11/1/90 Atlantis	4/91 Atlantis
IML-01	n.a. n.a.	n.a. n.a.	4/20/90 Discovery	4/11/90 Columbia	12/12/90 Columbia	12/91 Atlantis
TDRS E	n.a. n.a.	n.a. n.a.	n.a. n.a.	11/18/90 Discovery	1/31/91 Discovery	7/91 Discovery
ATLAS-01	n.a. n.a.	n.a. n.a.	1/17/91 Discovery	12/20/90 Columbia	4/4/91 Columbia	4/92 Atlantis
TSS-01	12/22/87 Atlantis	n.a. n.a.	10/25/90 Columbia	1/31/91 Atlantis	5/16/91 Discovery	3/92 Discovery
UARS	n.a. n.a.	n.a. n.a.	n.a. n.a.	9/26/91 Discovery	8/22/91 Discovery	11/91 Discovery

especially important for the Space Station assembly because the orbiters are not equally capable. Thus, station payloads should be designed to go on any orbiter, and perhaps to be assembled in various sequences where possible.

Of course some of the rescheduling shown is due to payload-specific problems and the Challenger accident, but that is the nature of a manned, reusable system: Because it is manned, safety comes first and delays result. Because it is reusable, delays in one mission necessarily delay later missions using the same orbiter. Thus, predicted flight rates are already less than a fourth of those predicted at the program's inception.

Table 14.3. Launch Vehicle Success Rate Through 1988

Vehicle	Average Success Rate
Ariane	77.8%
Atlas E	89.3%
Atlas/Centaur	89.7%
Titan	94.4%
Delta	95.2%
Shuttle	96%
Scout	100%
Saturn V	100%

Table 14.4. Major Manned Space Program Costs

Program	Cost
Apollo	$85 billion[36]
Mercury, Gemini, Apollo, Skylab, and Apollo-Soyez	$100 billion[37]
Shuttle (through 1990)	$65 billion

The performance of the Shuttle may also be measured against other launch vehicle programs. However, this is difficult because of the uniqueness of the Shuttle. Table 14.3 shows launch success rates for various vehicles through 1988, which must be read in recognition that the Shuttle is the only man-rated vehicle launched by the U.S. in the last 15 years.[38] The Shuttle compares well.

Table 14.4 compares the costs of three manned space programs. These comparisons are
meaningless without relevance to program achievements. Clearly the goal of exploration is defined broadly enough that the Shuttle program meets it. However, it is not so clear that the Shuttle meets the criteria of being the "next logical step", e.g., towards a mission to Mars. Program goals should be specific enough and achievable in short enough time that progress towards them can be definitively measured. The Apollo program certainly achieved its principal goal -- that of demonstrating U.S. technical preeminence by putting a man on the

moon and returning him safely to earth within 8 years of program inception.

Another simple, quantifiable measure of performance is payload mass capability. The original promise for the Space Shuttle was 65,000 pounds to a 180 km, 28 degree orbit.[39] Now, the average capability of the three active orbiters is about 49,000 pounds to the same orbit, a decrease of about 25%.[40] It should be noted here that the Shuttle puts about 250,000 pounds into orbit each launch, but about 200,000 pounds — the weight of the orbiter — comes back to Earth. This is a cost of reusability and of having a manned transportation vehicle. Another such cost arises from the need to make manned flights as safe as possible: Thus even unmanned payloads must also be man-rated. For example, the Shuttle-Centaur upper stage rocket was canceled when, after spending hundreds of millions of dollars, it was decided that the stage could not be made safe enough for deployment from the Shuttle.

The fact that the Shuttle gives us a manned space program is a clear, if non-quantitative measure of performance. Seen in this light much of the cost of the program can be attributed to the symbolic, rather than operational, aspects of the program. In this case it would be in the best interests of the nation to separate from the manned program that which does not have to be included. That is, payloads not needing manned intervention could be flown on ELVs, thus avoiding costly delays. In parallel the U.S. could pursue a vigorous, but independent, manned program where costs could be directly attributed to the symbolic. This separation could be done within NASA.

Given the Shuttle, and that its performance is expected to be in the range outlined above, it is possible to set goals for it and other programs that are consistent with relevant experience. For example, we should not expect the Shuttle to fly more than about 8 times a year, nor for less than about 4 billion dollars per year. While it is possible that the Shuttle may exceed these expectations, it is also possible that it will fail to meet them. We should not plan a space station needing more flights, nor reliable launch dates. Similarly policymakers should not commit our space program to planned capabilities of a new launch system until there is some evidence that performance will approximate promises.

Table 14.5. Summary comparison of goals established at program approval and actual achievements for the Shuttle program through 1990.

GOAL	PROMISE	PERFORMANCE
First Flight	1978	1981
Total cost	$51 billion	$65 billion
Average cost per flight:		
Including development	$88 million	$1.7 billion
Excluding development	$14 million	$1.1 billion
Flight rate:		
Annual average	48	4
Total	580	37
Payload mass	65,000 pounds	49,000 pounds
Manned capability	yes	yes
Reusable	yes	yes

Concluding Remarks

As the nation considers whether to embark on another major space program -- a human visit to Mars -- it is well to review the performance of the Shuttle program:

First, table 14.5 shows that the Shuttle has not achieved its goals. This should be remembered when considering the promises made for current civil space program alternatives.

Second, for useful program appraisal clearly-defined short-term goals are necessary. These can be used as measures of progress before large amounts of resources are committed. Goals should be specific and must include near-term objectives so that performance can be easily measured, and so that the program's political support can be solidified on a basis of cumulative success rather than on the hope of achieving fabulous promises in the distant future. That is, the evaluation of program performance with respect to specific goals should have quick turnaround times so that successes can be fed back for a positive effect on mission justification. Failures provide an opportunity for constructive correction of both policy direction and pace. Currently, many programs are designed with objectives that cannot be measured until a large amount of resources are spent. This complicates the evaluation process until it is too late to redirect monies from poorly-performing or non-performing projects. One possible set of criteria for program evaluation are cost, schedule, and performance as used in this paper.

Third, for the remaining half of its productive life we should operate the Shuttle based on a realistic view of its capabilities. During the 10 years of operations the Shuttle has proven to be neither cheap nor reliable; rather it is expensive, difficult to hold to a schedule, and the only U.S. means to launch humans into space. Acceptance of these facts would allow formulation of realistic launch policy. This, in turn, would allow for a realistic assessment of the Shuttle's role in the Space Station program and in any large scale human exploration program. Furthermore, when the decision to build a new launch system is made, it will be based on a clear picture of what shuttle capabilities allow, giving the new system a better chance of performing as promised.

Finally, for the U.S. to have a vigorous and productive manned space program we need to tailor our engineering, resources, and policy to fit realistic, achievable goals rather than being forced to alter unrealistic goals to fit the unforgiving realities of engineering, resources, and politics.[41]

Notes

1. Statement of James C. Fletcher, NASA Administrator, testimony before the Subcommittee on Space Science, and Applications, Committee on Science, Space, and Technology, House of Representatives, February 2, 1989.

2. J. Lawrence, *NASA Activities*, Vol. 21, #6, p.3 (1990).

3. Proceedings of the *Fourth National Space Symposium*. For example note J. Fletcher, "To be sure, the Shuttle has done what it was meant to do; it remains the most versatile, flexible, and useful flying machine in the world." (U.S. Space Foundation, Colorado Springs, CO) 1988.

4. "Report of the committee on the future of the U.S. space program," Washington, D.C. (1990).

5. The agreement can be defined as follows: All parties involved publicly agree upon goals for the program (in this case cost schedule, and capability), and this is established in Congressional Committee hearings and legislative reports which form the foundation on which public law is based. For comparison, goals which are not formally agreed upon, such as institutional needs or moving dollars to Congressional districts, may be thought of as the effective goals of the program. These goals are generally not part of the formal agreement.

6. The term "promise" has been chosen because it reflects a formal agreement, as described above. Some may assert that initial program promises are not to be taken seriously, i.e., they are overly-optimistic forecasts necessary to gain program approval. If program performance is important then a measure of performance is necessary.

Moreover, unless programs perform comparable to the expectations on which they are bought and sold, there can be little hope for any improvements in the policy process.

7. All cost data in this paper are in 1990 dollars, calculated using GNP price deflators from the Department of Commerce. Also total, annual, and average per-flight costs are calculated using NASA costs only, i.e., DoD costs are not included. Source for NASA budget data: 1970-1992 NASA Budget Estimates. (Actual expenditures used when available).

8. There are several analyses of Shuttle costs. Two representative examples are: Toman, M. A., and Macauley, M. K., No Free Launch: Efficient Space Transportation Pricing, *Land Economies*, Vol. 65, No. 2, May 1989, pp. 91-99 and "Space Shuttle Pricing Options", Congressional Budget Office, (March, 1985).

9. This paper does not attempt or perform an economic analysis of marginal costs for two reasons. First, Toman and Macauley (1989) have already tackled this problem. Their analysis was technical and because Shuttle program costs are not accounted in a way that allows understanding of how costs should be allocated, (i.e., which costs should be amortized over how many flights), they were forced to make several limiting assumptions. Until better data are available, further analysis of this type may not be useful. Second, it may not make sense (beyond what Toman and Macauley have done) to treat the Shuttle program in a standard economic manner: The de facto policy has been to fly the Shuttle as often as possible, and all cost data have originated under this policy. Furthermore, there is no economic "market" for Shuttle flights because most carry captive NASA payloads which fly on the Shuttle without consideration of price. For example, a NASA science payload can be given a "free" launch on the Shuttle or has the alternative of buying an ELV launch on the market. The science program sees its budget choices as: 1) no cost for the Shuttle launch and 2) a significant cost, i.e., market price, for an ELV launch. In this sense the Shuttle launch is "free" to the internal NASA users, although not free to taxpayers. Finally, resources are made available to the program through a political appropriations process not through a process of sales generating revenues.

10. For example see J. Logsdon, Science, Vol. 232, 1099, (1986); S. Pace, "Engineering Design and Political Choice: The Space Shuttle, 1969-1972", M.S. Thesis, MIT May, 1972; A. Roland, "The Shuttle: Triumph or Turkey", *Discover*, November, 1985 pp. 29-49; and R. D. Launius, "The Development of the Space Shuttle, 1967-1972: Technological Innovation and Governmental Politics", unpublished manuscript, NASA History Division, Washington, DC (1991).

11. NASA, Space Shuttle, information booklet, (October, 1972), p. 14.

12. As will be seen below, civil service salaries make up about 15% of the shuttle program's total cost. $8 billion errs on the side of being too large i.e., it makes the comparison more favorable to the original estimate. Adding this here gives a number which can be compared with our tabulation of actual total costs below.

13. NASA, Space Shuttle, information booklet, (October, 1972), p. 14.

14. For a complete description of NASA budget categories see the annual NASA budget requests to Congress.

15. It is beyond the scope of this paper to break down the program's cost into component parts (which would be necessary for a marginal cost calculation). Therefore the R&D /operations distinction is irrelevant, hence it has not been made.

16. All input necessary to compute the algorithm is available in the annual NASA budget request.

17. U.S. House of Representatives, United States Civilian Space Programs: 1958-1978 - Volume I, (January 1981) pp. 603-604.

18. NASA, Budget Estimates, Fiscal Year 1992, Volume 1. NASA, 1991. page 5F SUM 4.

19. U.S. General Accounting Office, *Budget Reimbursements: The National Aeronautics and Space Administration's Reimbursable Work*, GAO/NSIAD-87-171FS, U.S. G.A.O., June 1987, p. 9, and T. Dawson, personal communication.

20. The 1987 NASA budget also included an additional $2.3 billion (1990 $) for the orbiter which was procured to replace the Challenger. This figure is included in the total cost calculation.

21. U.S. General Accounting Office, Testimony presented by C.A. Bowsher, before Subcommittee on Government Activities and Transportation, House Committee on Government Operations, May 1, 1991. Additive costs are the variable costs of consumables associated with each flight, i.e., the costs of fuel, expendable hardware, etc..

22. The average costs per flight displayed in these two figures include shuttle development costs. This is so that projected average costs may be compared with original promises. To calculate an average cost per flight for the remainder of the program, divide average annual appropriation by average flight rate.

23. It should be noted that this number is marginal, as well as average, if annual flight rates and appropriations do not vary a great deal from these numbers.

24. The 1987 budget contained $2.1 billion for the replacement orbiter which would be $2.3 billion in 1990 dollars. To be conservative some additional costs were not counted: Approximately $400 million for structural spares had already been procured and were used in the replacement. Further, about $100 million for main engines was not included in the $2.3 billion.

25. Office of Technology Assessment *Access to Space: The Future of U.S. Space Transportation Systems* (OTA-ISC-415, Washington, DC: U.S. GPO) 1990.

26. This figure is arrived at by multiplying the expected chance of failure on one flight (0.02) by the cost of a replacement orbiter ($2.3 billion). For example, had the cost of Endeavor, the Challenger replacement, been assessed on a per flight "surcharge" basis rather than in a lump sum, it would be paid for after the fiftieth flight. Through 1990 the shuttle launched 38 times with one loss. Due to the fact that shuttle reliability can not be known with a high degree of accuracy, such a method of determining the surcharge amount might or might not recover the full cost of replacement orbiters. The $2.3 billion cost for replacement orbiters is conservative because it does not include costs of "structural spares" and shuttle main engines which were paid for elsewhere. See note 24.

27. For more information on Space Shuttle launch rates consult: Committee on NASA Scientific and Technical Program Reviews, Commission on Engineering and Technical Systems: *Assessment of Constraints on Space Shuttle Launch Rates* (1983) and National Research Council *Post-Challenger Assessment of Space Shuttle Flight Rates and Utilization* (National Academy Press, Washington, DC. 1986).

28. National Research Council (1986), pp. 4-5 (see note 27).

29. Atlantis flew twice before the Challenger accident and gave indications that it could be expected to perform as well as Discovery.

30. Not surprisingly, many small problems are discovered in preparation for a first flight. Similarly, the seventh flight of Columbia is ignored as it had been out of service while at the factory for several years for modifications.

31. However a 1988 OTA report has expressed concerns that the 75 workday turn-around is too ambitious. See *Reducing Launch Operations Costs: New Technologies and Practices*, OTA-TM-ISC-28 (Washington, DC: U.S. GPO, 1988).

32. U.S. Congress, Office of Technology Assessment, Reducing Launch Operations Costs: New Technologies and Practices, OTA-TM-ISC-28 (Washington, DC: U.S. GPO, September, 1988). p. 41

33. Additionally, in an unpublished 1986 white paper prepared for the Rogers' Commission NASA determined that costs per flight were underestimated by 475%.

34. Even more schedule predictions were made than could be presented on the graph. This graph has been provided by Bruce Murray.

35. For example, flight 20 was predicted (in July, 1979) to occur early in 1982. Given satellite lead times, construction of the payload had to be occurring by July, 1979, i.e., costs incurred. The chart shows that flight 20 actually launched in mid-1985, three years late. If it were a communications satellite the builder would have paid three years interest on the investment -- some millions of dollars.

36. This number is probably underestimated. The Augustine committee cites $95 billion for the cost of Apollo, source: NASA Historical Data Book, Volume II: Programs and Projects 1958-1968, (NASA SP 4012,) (1988). NASA, Washington, D.C.

37. NASA (1988) p. 121 (see note 37).

38. Chart reproduced from *Reducing Launch Operations Costs*, Office of Technology Assessment (1988). Data on Saturn V from *U. S. Civilian Space Programs 1958-1978*, Committee on Science and Technology, U.S. House of Representatives (1981).

39. NASA, Space Shuttle, information booklet. (October, 1972), p.1.

40. *Access to Space: The Future of U.S. Space Transportation Systems.* p.41, and T. Dawson personal communication. It should be noted that some NASA payload weights include items such as the space suits, the robot arm, and the galley, all of which return to Earth with the orbiter.

41. We would like to acknowledge the useful comments made by Ronald Brunner, Terry Dawson, Pete Didisheim, Millard J. Habegger, Donald Hearth, Molly Macauley, Richard McCray, David Moore, Richard Obermann, Michael Rodemeyer, William Smith, Alan Stern, Skip Stiles and Albert Wheelon on earlier drafts of this paper. We would also like to thank Patricia Duensing for her efforts in the preparation of this paper. However, as usual, all responsibility for the content lies solely with the authors.

Part Four
CONCLUSIONS

This part consists of only two essays which deliver a very similar message from two very different points of view. Limerick, as an historian of the American West, recognizes the importance of myths, and based on her familiarity with the myth of the frontier argues with some humor that it is misused in the way NASA applies it to space exploration. She argues that NASA needs an updated frontier myth.

Byerly's chapter concludes the book but is not a summary. Rather he attempts to find the overall lesson or message of the previous chapters. Not surprisingly he finds that the times require a new approach and he begins to suggest what it might be. His message is congruent with Limerick's but is delivered in a very different style, from a very different point of view.

Chapter 15
IMAGINED FRONTIERS: WESTWARD EXPANSION AND THE FUTURE OF THE SPACE PROGRAM

Patricia Nelson Limerick

A NATION'S PREOCCUPATION WITH HISTORY IS NOT INFREQUENTLY AN EFFORT TO OBTAIN A PASSPORT TO THE FUTURE. OFTEN IT IS A FORGED PASSPORT.
—HOFFER

There is an old Western joke about a mild-mannered, meek stranger who was having a drink in a tough Western saloon, when a black-hatted bad guy came into the saloon, and began crowding people, and knocking over their drinks. After a few minutes of this, the mild-mannered, meek stranger went over to the bully, and said, "I'm giving you five minutes to get out of town." To everyone's amazement, the bully immediately packed up and left and got out of town. The very-impressed townspeople crowded around the mild-mannered stranger, and asked him the obvious question: "What *would* you have done if he hadn't left in five minutes?' and the mild-mannered stranger said, "I believe I would have extended the time."

I open with this story because the backers of the space program and I have been heading for a showdown for a while, and now that it is High Noon, Main Street, 1991, and the showdown is ready to take place, I am myself somewhat inclined to extend the time, to ask if we could do this in 1992, or perhaps 1993. Exposing the mistakes and misleading assumptions in other people's articles of faith is only pleasure to a certain kind of bully; for the rest of us mild-mannered folk, "extending the time" seems a more desirable option than plunging in to shed a bright light on other people's illusions.

I am not, in any case, an historian of the space program. I have read a bit in the field, but I am still learning terms and events and people's names that are very familiar to most readers of this volume. And when it comes to reasons for extending the scheduled time for our showdown, it weighs on me that technology is hardly my strong suit.

If the space program had stayed in the terrain of engineering, you would clearly not have me writing here, because engineering is not my business. But the history of the American West *is* my business, and at different stages of its history, the promoters of the space program have left the terrain of the engineer, and jumped into the terrain of the Western historian. I refer, of course, to the common pattern of characterizing space as the "new frontier," but let me make it clear that this was not my idea. If I had been around at the start of this whole business, and if someone had asked me if I thought NASA would be well-advised to compare space to the frontier, I probably would have advised against it. "Don't you have enough cans of worms?" I would have asked. "Why borrow ours?"

But nobody asked me — indeed, neither I nor anyone like me was there to be asked, and here we are, with three decades of speeches and reports and P.R. campaigns now accumulated, tying the space frontier to mine. So we are related, we are in-laws, and that is why we are having this showdown. Even though I would not have recommended the relationship, I think now we ought to stick by it — because I think it would do the space program a world of good to *keep* the frontier metaphor, but — for the first time — *to take it seriously*. And I think it would do Western historians a world of good to have this compelling public demonstration of our relevance and our value to society.

Before going any further, I have to get my operating premise on record: The metaphors and comparisons and analogies that a group chooses do in fact carry a lot of meaning, and can indeed control actual behavior. The metaphor you choose guides your decisions — it makes some alternatives seem logical and necessary, while it makes other alternatives nearly invisible. My point, then, is that the pattern of comparing space to the frontier is not a light or trivial matter — that, in other words, thought, behavior, and especially appraisal of what options are available, are all limited by a misused metaphor. And on the other side, the space community's thinking, and sense of options and alternatives, could gain new force and new *range*, with a properly used metaphor. In fact, I present this paper as the work of a Metaphorical Engineer — since the engineering of metaphors may be what I do best and since only the furthest stretching of a metaphor could cast me as an engineer.

In case there is some doubt as to what pattern of thought I'm addressing, consider President Ronald Reagan's speech, delivered on the Fourth of July, 1982, at the landing of the shuttle Columbia:

> The quest of new frontiers for the betterment of our homes and families is a crucial part of our national character...The pioneer spirit still flourishes in America.
>
> In the future, as in the past, our freedom, independence, and national well-being will be tied to new achievements, new discoveries and pushing back frontiers. The fourth landing of the Columbia is the historical equivalent to the driving of the golden spike which completed the first transcontinental railroad.

In making this comparison, neither Reagan nor his speech writers had to *think*; by 1982, this way of speaking and thinking was so well set that no one would say, "Does that comparison make any sense? Does the landing of the space shuttle really have anything to do with the building of the transcontinental railroad?"

The first thing that strikes the Western historian is that President Reagan and his speech writers thought that this allusion to the frontier was a happy comparison, the right comparison for an occasion of congratulations. But add a few facts about the picture of the Golden Spike, and things look a bit different. What the President thought was just a light "Have a Nice Day" reference to history *could* have been a pretty useful warning, *if* anyone had taken it seriously.

When they connected the railroad lines at Promontory Point in 1869, the representative from the Central Pacific, Leland Sanford, proved unfamiliar with a sledgehammer, and could not hit the Golden Spike. The inability of a railroad executive to perform the most elemental act of railroad construction might — if anyone wanted to take these analogies seriously — say something about the gap between executive planning and hands-on implementation that transportation industries are vulnerable to, and it would deepen that point to recognize that much of the railroad trackage theoretically "completed" in 1869 actually had been laid in such a rush that much of it had to be laid again almost immediately. In other words, a reference to the Golden Spike, to anyone who is serious about history, is also a reference to enterprises done with too much haste and grandstanding, and with too little care for detail.

Ronald Reagan also did not know, or care, that one half of the first transcontinental, the Union Pacific Railroad, went bankrupt twenty-five years later in the depression of the 1890's, or that the other half, the Central Pacific, even though it became more prosperous, did so by keeping a stranglehold on Pacific Coast traffic, charging all that the traffic would bear, through its affiliate the Southern Pacific, the company known as the Octopus, the company whose chief attorney was widely understood to hold much greater power in the state of California than the so-called governor did. With all that prosperity, the Central Pacific still played out a prolonged drama in trying to avoid paying back its government loans, and in trying to get out of the interest payments.

Add to this the far-reaching corruption in Congress that came out of federal aid to railroads, and add the rough and even brutal working conditions on the railroad, especially for the Chinese working on the Central Pacific in the Sierras in winter (we didn't bother to keep track of how many died, the construction manager said later; we knew we could replace them); add it all together — executive misbehavior, large scale corruption, shoddy construction, brutal labor exploitation, financial inefficiency — and when the President compared the shuttle landing to the Golden Spike, it's a wonder that someone from NASA didn't hit him, to defend the agency's honor. It's a wonder no one — no shuttle pilot, mission coordinator, mechanic, or technician — said, "Now cut that out — we may have our problems, but it's nowhere near that bad."

That's the joy of the present status of the frontier metaphor — you can use it to say things that are really quite insulting to the integrity of the space program, and the people thus insulted will smile and say, "Thank you."

But think of the *value* that's hidden in the metaphor. If you take it seriously, then it can be something more than a thoughtless compliment; it can be a veritable checklist of problems to watch out for in enterprises involving large scale transportation and enormous federal funding.

So there, with the Golden Spike, we have one exercise in taking historical analogies seriously, but before we go further, we should probably address the thought that this is certainly not the Western history one learns from movies, and it is not even the Western history one learns in school. And that brings me to a not-very-widely-recognized fact: in the same era that saw the rise of the space program, there has been a revolution in the writing of the history of Western expansion. If you covered Western history in school, you probably got a dose of the old frontier school of historiography, the school inaugurated by the historian Frederick Jackson Turner in 1893, with his famous speech, "The Significance of the Frontier in American History."

In such "Old Hat" frontier history, heroic white pioneers brought civilization to a savage wilderness; they put the virgin lands of the West to a higher economic use; they forged democracy in their simple, equal pioneer communities; they fulfilled the nations' destiny and they made their own fortunes; and then, when the process was complete, when the wilderness was civilized and nature was improved into farms and cities, then, around 1890, the frontier closed.

To put it in a nutshell — we don't believe that anymore. Parts of the picture now look downright wrong — American democracy came from thinkers on the East Coast, not from humble settlements in the interior; the unsettled, virgin wilderness was actually the home of Indian people and Hispanic people; and, in episodes like mining rushes, far more people failed than succeeded, and those who succeeded did so by building large companies that employed others as wage workers under conditions that were a lot closer to complex industrial life than to simple pioneer democracy.

In other words, the timing is crucial here, with the development of the space program and the development of a New Western History running neck and neck. That presents us with an opportunity to see if an application of the New Western History might aid in clarifying space policy thinking.

Remember, I never pushed the frontier metaphor on anyone; I never asked engineers to compare themselves to Kit Carson, or George Armstrong Custer. They did it on their own — they came into my territory, picked up the concept of the frontier and westward expansion, took it back to their turf, and made it a key item in their package of promotional tools. They used it on Congress, they used it on Presidents, they used it on the public, and — this is what most concerns me — they used it on themselves.

My concern isn't that the space program used the idea as a cynical sales pitch; my concern is that they used it as a *sincere* sales pitch — that they sold it to themselves. This is where I switch from metaphorical engineer to metaphorical consumer fraud watchperson: when the space program supporters bought their version of the frontier, they bought a very dated, outmoded, and even dangerous model, the kind of model that's old enough to explode unexpectedly, the kind of model that can't bear the weight of serious *use*. And that's the center of my consumer warning: the model of westward expansion the space people bought comes *without consequences*, and without allowances for failure.

In the old frontier model, everything ends in 1890, with the continent settled and tamed and improved, and it is in its essence a *happy* ending – with no room in it for the many failures that make my region today conspicuous for its ghost towns. Not only is the history of westward expansion full of examples of failures – in particular, of short-cuts that turned into death-traps – it is also full of unresolved consequences: of minesites leaching toxic chemicals into streams and rivers, of demoralized conquered people trapped by alcoholism and unemployment on reservations, of over-allocated streamflow and depleted groundwater, of periodic fires in forest lands that only the silliest of optimists would call "managed". The new, real Western history is full of consequences and instructive failures; the old, false frontier history denied consequences and overruled failure. And that is why this seems so serious to me – denial of consequence and evasion of failure adds up to *exactly* the wrong prescription for the space program. The old metaphor is a prescription for complacency, and complacency is the perfect soil for error, haste, and a kind of compulsion that masquerades as free choice.

In other words, the space program needed a metaphor that would keep people alert, regularly examining their own behavior and their own thinking, and instead they got a metaphor with exactly the opposite properties – a metaphor that makes its believers complacent, even smug, and inattentive to their own operating assumptions. It could not be more dramatic if a patient who needed a stimulant got a sedative instead, or if a person using over-the-counter medicines misread the label, and thought that it said that you *must always* take a drowsiness-inducing antihistamine if you intend to operate heavy equipment.

To see people operating heavy equipment while under the influence of a metaphor with ten or twenty times the stupor-inducing wallop of a standard antihistamine, look at the Paine Commission Report of 1986. If you are planning the next fifty years of the space program, then that is heavy equipment indeed, and super-alertness would seem to be called for. But the very title of the report swings us away from alertness, and back to complacency: *Pioneering the Space Frontier*, they called it, and I can assure you that this is not a reference to the exciting new perspectives of recent Western history. The Paine Commission's Western history is vintage 1890's; it is Turnerian frontier history, with a fervent optimism and cheeriness that might well have made Frederick Jackson Turner himself a bit ill-at-ease. Their money for consultants must have gone entirely to engineers and physicists, one assumes, leaving not even a modest *per diem* for an historian to tell them that their plan for the 21st century wasn't yet in the 20th century when it came to Western history.

The problem is clearest in their opening "Rationale for Exploring and Settling the Solar System": "Five centuries after Columbus opened access to 'the New World,'" they begin, and the historian is already full of questions. Do they know that the New World had natives? That, in a considerably more fundamental way, the emigration of people from Asia 10,000 to 30,000 years ago "opened access" to the New World? Do they know that Columbus' career was an almost immediate mess — the island "settlers" found hardly any gold; the Indians were soon dying from disease and forced labor; the colonists were soon leading mutinies against Columbus; Columbus' three last voyages were disappointments and even disasters, with him returning in chains from one of them?

When, in other words, the Paine Commission suggests that we follow in Columbus' footsteps, do they have *even a clue* as to where those footsteps went?

Then move to the next sentence: "The promise of virgin land and the opportunity to live in freedom brought our ancestors to the shores of North America." Ask "What's wrong with this picture?" and the answers are endless. The lands were not "virgin" (whatever that interesting word means); Indians hunted, farmed, burned and possessed them. "Our" ancestors — and there is a Tonto-and-the-Lone-Ranger question here, "What do you mean *we*, white man?" — our ancestors came for a whole hodgepodge of reasons. Many of them came for quick short-term profit. (Consider, for instance, the strange and unsettling behavior of the original settlers of Jamestown, Virginia in 1607, who thought they were surely entitled to find gold and riches equivalent to what the Spanish found in Mexico, and who were therefore so determined to be fortune-hunters and fortune-finders that they simply refused to farm for the first years, preferring to steal corn from the Indians, or even to starve.)

A whole other set of "our ancestors" came — not for "the opportunity to live in freedom," but because they were slaves, and it is in this denial of the fact of slavery that the Paine Commission looks mostly wildly out of touch with reality. Beyond race, many white people came as indentured servants, and many of those died before they could earn their way to freedom.

This is "a species destined to expand," this first item in the Rationale concludes, and here the historian simply wants to say, "Watch out for this Manifest Destiny business; it is a lot trickier than it looks. Commit yourself to a *destiny*, and you are handing over your free will; you are *volunteering* for compulsion; you are doing things because you *must* do them, not because you have reflected, pondered, and chosen to do them."

When we turn to the second item in the rationale, the muddle gets even deeper. "The settlement of North America and other continents was a prelude to humanity's greatest challenge: the space frontier." That sentence, by itself, I would be willing to let stand, because the conclusion that seems to come from it is such a sensible one: if one frontier is a prelude to the other, *then learn all you can about the first one* before you leap into the second one. But that's not the conclusion of the Commissioners — their paragraph ends quite differently. "As we develop new lands... we must stimulate individual initiative and free enterprise in space." Well, maybe ... but if that's what we *must* do, then it is our obligation to make this as well-informed a move as possible. That is, we must know as much as we can about what "individual initiative and free enterprise" achieved in the American West. While that will certainly include the fortunes made and the jobs created, it will also include the ghost towns and the floods, the Dust Bowl and the Exxon Valdez, plus the clashes between users — the perennial squabble, for instance, between the hydraulic miners and the downstream farmers who did not consider mining rubble added to the streamflow to be an improvement. On this part of the Rationale, in other words, the space planners give us a picture of harmony and progress where historical reality shows us something closer to a muddle, with good news and bad news thoroughly mixed.

Move to the third item, and you find this: "Historically, wealth has been created when the power of the human intellect combined abundant energy with rich material resources. Now America can create new wealth on the space frontier ..."

Well, maybe ... but Western history carries other lessons. A mine, it was often said, is a hole in the ground into which a fool drops his money, and Western historians have looked at particular mining areas like Nevada's Comstock Lode, and concluded that investments were greater than returns. That is, when you total up the wealth invested in equipment and excavation and labor and transportation and processing and stock speculation, you get a total greater than the wealth realized in silver or gold. Now this is not a law of nature;

things do not *have* to be this way; but it is still a good reason to raise a cynical eyebrow when someone tells you that a new frontier is going to create "new wealth."

One could go on, item by item, citing, for instance, the Commissioners' momentary contact with Western reality when they recognize the central role of the federal government in developing Western transportation – a momentary contact with reality that ends almost immediately *before* the historically established risk of corruption, financial inefficiency, debt, dependence, and the irritability that comes with dependence, can be mentioned.

Juxtapose their final resolution – "We must remain true to our values as Americans: to go forward peacefully and to respect the integrity of planetary bodies and alien life forms, with equality of opportunity for all" – to the reality of Indian devastation by disease, alcohol, loss of territory, and coercive attempts at assimilation; put those two things together – the lofty good intention, the tragic reality of the past – and one almost feels obligated to take up the mission of warning the aliens, telling them to keep their many eyes on their wallets, when they hear these admirable intentions invoked. Space aliens are usually not very sympathetic characters in novels or movies, but this particular reference to frontier history makes one feel definite twinges of compassion for them, regardless of how many eyes, arms, or antennae they prove to have. To put this more directly, if we are going to find infinite resources out there and "create wealth," and find homes and jobs for millions of humans, then this cannot be good news for the aliens or the planetary bodies, who are going to absorb some losses while getting their dose of equal opportunity for all.

My point, with all these items, is that the Paine Commissioners and the other space advocates have trivialized this whole frontier business, and made it just silly, when it could be instructive and maybe even inspirational. And the ways in which they have trivialized Western history are exactly parallel to the ways in which they have trivialized their own enterprise, evading the reality of failure and denying the importance of consequences.

As I have tried to learn about the space program I have had many discussions with members of the community. They have translated acronyms[1] for me and explained broader issues to me. They believe strongly in space exploration and space science and are very concerned with NASA and with the future of their program. Our discussions seemed to have a common flavor. As a community there seemed to be a common behavior.

What metaphor would I use to characterize the commonality I saw? It was like a group of kinfolk gathered to discuss the troubled condition of a wealthy, powerful, but increasingly dysfunctional relative. The relative — let's make him an uncle — still has a lot of power to wield, from punishments to wealth, and maybe even good to do. But NASA, which is to say the uncle, is showing bizarre behavior — "unreliable" several said — and acting under a number of delusions. And the kinfolk are at present paralyzed; beyond expressing their concern to each other, nobody knows what to do.

To the newcomer to this scene, the solution seems self-evident at first. Since NASA's present behavior is seriously jeopardizing the program's longterm future, then someone must tell NASA that — and fast. But then, as the newcomer observes a bit longer, the arguments against full disclosure appear.

You can't speak frankly to this aging agency — because this agency can be vindictive, punishing dependents for their honesty; because this agency is enormously thin skinned, flying off the handle at a word of criticism, taking even well-intentioned appraisals as "NASA-bashing"; and finally, because this agency is in the grip of irrational convictions and passions, and thereby not in the mood for conversation.

The word "frontier" is used in promotional statements, but another word, also full of historical associations, is absolutely central to the whole space culture. That is the word "mission." This is a word that the community takes for granted; members do not cross themselves, nor genuflect, when they use it: To their ears, mission probably sounds like task, or assignment, or project. But it's an interesting and distinctive word, and it must have been chosen in the first place — and fastened onto — because of its connotations. Unlike a task or activity, a mission has a higher purpose; it serves higher powers, it carries its own justification, and only the heathen ask, "Why? What's the point?"

So that is one part of why no one can engage NASA in rational discussion over its problems — NASA administrators and staff are, in practice, missionaries, even though they don't call themselves that (but *what else* can you call people who *must* have *missions* and bigger and better missions at that?). Think of them as missionaries, and NASA people appear as what they are — creatures of a stiff and brittle faith, warding off infidels, and never understanding why the heathen don't have the sense to convert.

This institutional failure to communicate, this reluctance to communicate, this hostility to communications; that is the most troubling impression I've gotten from my discussions, most troubling because it is so reminiscent of the failed communication Richard Feynman described in his reflections on the Challenger disaster[2], and because this shutting down of conversation, with the expletive "NASA-bashing," certifies that this is an agency *no one can help*. If someone doesn't figure out this one, if someone can't get a message through to NASA that this defensiveness, this complete unwillingness to face open appraisal, is self-destructive and deeply shortsighted, then I have to think the party's over, and I would then shift to a "people at a wake" metaphor.

Let me turn to the other problem of communication that's come up several times as I talk with space people – the problem of promotional institutions that come to believe their own rhetoric, salesmen that end up buying their own salespitch. It is clear to me that NASA – like all federal agencies – has to put on regular theatrical performances. "Success-orientation," it's been called, or under-estimating budgets, or chanting the ritual line, "We don't see any problems." I'm oddly reluctant to call that lying or hypocrisy, because I think enough performances, enough repetitions, and the actors cease to be able to tell that they're acting. What NASA's example proves is the proposition that it is difficult for salesmen to sell programs without at last selling themselves – and I mean that in both senses, as in selling themselves the package, and as in selling themselves to the devil.

But is the sale permanent? Must this system of deception and self-deception go on forever? More important, *can* it go on forever? Or will the Gramm-Rudman budget-balancing days require a new adaptation – where budgets have to become realistic and honest statements, not works of theatrical art? Can NASA make that adaptation? My impression is "probably not," but who knows? NASA's done a lot of remarkable things, and this change could be the next one.

And that – NASA's catalogue of genuine achievements – brings me to my last point, which is that the central problem may be this: NASA is aging, and it is an institution whose aging process is utterly without grace. Anyone who saw the documentary on PBS, "The Other Side of the Moon," on the Apollo astronauts twenty years later, got a concentrated dose of this point. It is in some ways more surprising than seeing old baseball players, or old football players, to see astronauts, who once embodied youthful vigor, with receding hairlines, triple divorces, and oddly aimless careers.

NASA sprang to life, leapt into action, raced to the moon – and it is pretty tough to hit a phase of institutional life where all this springing and leaping and racing no longer works. In the institution's early years, a group of bright, very tough, very vigorous young men had their peak experiences together, and naturally they yearn for those times, and naturally they don't know what to do about the passage of time and the changing of circumstances.

Historians of European expansion know this pattern well; explorers, by and large, don't age well. Columbus, Cortez, Coronado, Meriwether Lewis, John C. Fremont, Clarence King — one can put together a long list of daring young men who arrived at middle age looking pretty bedraggled. They did not have the word or concept "enough" in their vocabulary, though they certainly had the word "more" — and they fell into the pattern that my students and I call "staying too long at Vegas," not knowing when to call in their bets, not noticing when their luck changed.

Aging explorers are notoriously clumsy when it comes to passing on the torch; once the first round of daring and impulsiveness is over, it is nearly impossible to find a formula that will balance caution and daring, to give up a romantic past and live in the humdrum present. And NASA has to undergo its midlife crisis in the cruelest times — in budget-cutting times. NASA, moreover, is at the same time absorbing the blow of having a history. By that, I mean that an organization that defined itself as young, fresh, free, innovative, is very slowly waking up to the recognition that it has a history, and having a history is like carrying around barnacles, or sticky wads of flypaper. Some of the comments I have heard indicate that the barnacles are pretty thick — that NASA is virtually encrusted with its habits, routines, conventions, and rituals; and this is a pretty rough thing to face up to in an organization that was supposed to embody innovation.

Up to this point, if we can shift back to the analogy of a gathering of concerned kinfolk of a powerful but troubled relative, my conclusion is that things look pretty bad. The patient is taking aging in the roughest, most graceless way; the patient can no longer distinguish sales pitches from deep convictions; the patient's capacity to injure himself is considerable; the patient alternates erratically between recklessness and timidity; and the absolute worst, the patient won't talk about it, and flies off the handle at a word of criticism, and insists, compulsively, that everything is fine.

All that is pretty bad, but there is another way of looking at things that is actually quite calming. That is, in its *actual*, rather than *perceived* dimensions, this is not much of a crisis at all. NASA still has gratifying public and congressional support, and it still has money. It's being asked to give up some glamour, some human drama, some arbitrary freedom of action, some illusions of omnipotence — to shift from missions to tasks, to live, in other words, on a plane of reality close to where most other people and organizations already live.

So NASA is being asked to grow up.

So big deal. Some crisis.

Turn the heat down a little on this situation, cut back a little on crisis-modality self-dramatization, add some perspective, and this is a pretty manageable transition. It doesn't have to be a crisis, even though NASA seems to want it to be one.

Now this is where I get nervous again, because I am a Western historian, not a space program historian, and not a policy planner, and certainly not a prophet. But I will continue to put myself forward as a Metaphorical Engineer, and as a Metaphorical Engineer, I am sure that if some resources are put into this, i.e., to develop a metaphor for the space program that *deepens*, rather than trivializes the enterprise, then I think a lot of other things will fall into place. If I were in the space community, I would be driven crazy by the repeated and endless demands that the space program develop a clear plan and a clear set of goals. That is simply not how a creative human enterprise proceeds — you don't know where you're going until you start going there; you start writing a book on ghost towns in Western America, and you end up writing about what's wrong with the space program, and that comes upon you more by surprise than because you outlined this goal for yourself ten years ago.

So this would be my advice to NASA: when people demand that you set forward your plans and goals in definite ten-year increments, smile patiently — you can even fill out their forms with platitudes if they insist — but put some real energy into getting a metaphor that you can trust, a metaphor that won't betray you into complacency or compulsion. When you consider the pool of applicants for that metaphor, let the real Western history be one of the candidates.

Notes

1. I wonder if Western historians would be more acceptable if we used acronyms. There are some first rate ones waiting to happen-take the transcontinental railroad, the Central Pacific and the Union Pacific, CPUP, or Seapup. Or Columbus's Nina, Pinta and Santa Maria, NPSM. Will I relate better if I call it the Nipsum Mission of 1492?

2. Feynman, R.P. "*What Do You Care What Other People Think?*", Norton, NY, 1988. See "Afterthoughts" on his role in the investigation of the *Challenger* accident, pp. 212-215.

Chapter 16
CAN THE UNITED STATES CONDUCT A VIGOROUS CIVILIAN SPACE PROGRAM?

Radford Byerly, Jr.

IT DOESN'T TAKE MONEY TO THINK!
THAT'S WHEN YOU DO YOUR BEST THINKING,
WHEN YOU HAVE NO MONEY.
—C. JOHNSON[1]

Problems

We have tried to do two things in the preceding chapters: First to show how our space program is significantly affected by the context in which it operates and by the governmental processes which are a large part of that context. The budget process described by Telson (and mentioned by others) is the leading example. The second thing we tried to do was to look ahead; to see where we should go, what policies we should adopt.

The preceding chapters — especially those by Fellows, Guasteferro, and Coleman — illuminate some of the incentives implicit in the current policies for operating our space program and in the policy context outside the program, and show how these incentives are counterproductive. For example the system of incentives seems to drive toward large, inflexible programs which, because of their size and lack of resilience, are likely to fail; that is, to perform significantly and substantially below the promised level.

It is unlikely that the external policy context will be changed in order to facilitate a space program — for example, the Gramm-Rudman budget process is driven by a deficit unlikely to disappear soon and so budgets and the budget process are likely to remain problematical. Thus changes will have to be made within the program in order to accommodate the external environment just as boosters are designed to overcome gravity. And just as engineers accept gravity as a given, the existence of an onerous policy context must be accepted and dealt with, not bewailed.

As for looking ahead, it is always difficult to know the future. This has two aspects: First, the policy context could change. Second, future preferences, i.e. what we want to do in space, could be different from what they are now.

Consider the first possibility, a different policy context in the future. This is not a difficult problem because a major characteristic of the present context is its unpredictability. Therefore learning how to operate flexibly, resiliently in the present policy context is good preparation for a range of future ones.

The second possibility bears more discussion. On the one hand it seems that there are changing fashions in what is desired of the space program. For example, the priority given to applications, especially commercial applications, seems to wax and wane. The early and outstanding success of commercial communications satellites leads many to believe that other such successes await discovery: Pots of gold somewhere over the rainbow, despite the fact that diligent searching has found none. Similarly there is now pressure to use space to address the problem of global change through the NASA Earth Observing System. Webster has described this program and its problems.

On the other hand there has been a remarkably steady vision of what the core of the program should be about: The exploration and understanding of outer space. That is to say, human exploration and scientific observation. Even the tension between advocates of these two goals has been fairly steady, with many scientists steadfastly denying that human spaceflight has any but the smallest scientific payoff (which may be true but irrelevant). Certainly the basic plan for manned exploration seems rather non-controversial. For example questions about whether or not to bypass the Moon on our way to Mars turn on efficacy rather than on any doubt of the ultimate goal. Explorers want to go to Mars next, the issue is how.

To some extent the commercialization issue is over "how" to conduct a space program, but the broader question of the importance of applications is not. It has to do with the purpose of a space program. That is, should we go into space for practical, earthly purposes or to pursue an extraplanetary vision? That question cannot be answered within the space program.

In a democracy, such questions ultimately have to go to the people, perhaps through their elected representatives. This writer's belief is that if such a question is clearly and fairly presented then a clear and fair answer will be forthcoming. (Thus the need for debate, as argued in the Introduction.) My further belief is that given a clear choice the people of the United States would clearly support a vigorous exploration program, although they would support neither unlimited funding for exploration nor the termination of science and applications programs.

The point is, there are three kinds of problems facing the space program today: First, problems within the program that can be fixed within the program, given the will to do so. This would include mission giantism and overreliance on Shuttle. Second, problems outside the program — such as the Federal deficit — that are unlikely to be addressed in terms of the space program: We will have to learn to manage with the deficit and its ramifications. Third, there are a few problems, such as uncertainty about the ultimate purpose of the space program, which will have to be solved outside the program, but whose solution can be catalyzed and optimized by the space program, e.g. by stimulating informed debate.

The System

The preceding chapters, taken together describe a way of doing business that is familiar and thus comfortable; it has been the pattern since Apollo. In this pattern costs are unimportant (contrary to mythology) and decisions are based on traditional and political considerations more than technical merit. Macauley (in this volume) has described some of the reasons for this behavior in terms of the principal-agent problem. (She also points out the importance of finding out what citizens want from their space program.)

In short, the Apollo paradigm reigns, as has been discussed in the Introduction. There have been few effective incentives to change because overall support for the program has been relatively strong and appropriations continue to roll in. Although as Pielke and Byerly show in their chapter recent program performance often has been far below program promises, that has not significantly affected the conduct of the program. The external world has been giving feedback to the program in many ways – from the Challenger accident to the near cancellation of the Space Station by Congress in the spring of 1991[2]. But any negative feedback has been heard as philistinian threat rather than as any sign that there might be internal problems. In her chapter Limerick analyzed this in terms of the false myth on which the program is based.

Prospects for Change

The space program needs a different set of driving incentives. The present set works fine to preserve the existing system and a program that is built on image and largesse. The system resists change: Congress will not make changes that threaten jobs; the Agency's first priority is self-preservation; the contractors are squeezed by reductions in defense work and are happy to get the 10-to-20 year jobs offered by giant programs; and the White House likes the glory of announcing a trip to Mars which distant future administrations will have to finance. (To its great credit the White-House-based National Space Council, chaired by the Vice President, has tried to insist on a reasonable approach to the Mars mission, although this may be in part due to consideration that some of the bills will have to be paid during a Quayle presidency.)

Are there then no prospects for constructive changes?

The coming budget crunch may be a well-disguised blessing. The program is in dire need of some fresh ideas, and as Caldwell Johnson has said, it doesn't cost money to think. He went on to say that you do your best thinking when there is no money, which leads to the speculative hypothesis that a surplus of money may lead to a deficit of thinking. Program history would seem to support that hypothesis, as does common sense. Real thinking and the subsequent implementation of new ideas is *hard work*. Much harder than securing appropriations and writing checks.

Telson and Macauley have outlined the general budget situation in their respective chapters. Clearly we are likely to face a virtually level space budget for several years. The Augustine committee has noted that the space program as planned is overambitious: "NASA is currently over committed in terms of program obligations relative to resources available – –in short it is trying to do too much ...."[3] Thus it is very likely that there will have to be a retrenchment, a scaling back of grand plans.

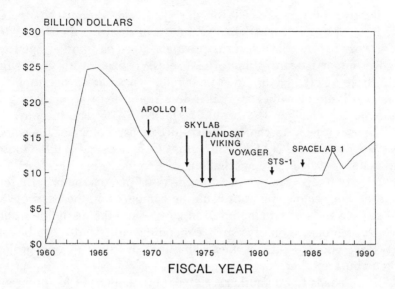

Figure 16.1: Past Major Accomplishments. The NASA budget (in FY 1992 dollars) showing major accomplishments in the mid-70 s and mid-80 s[4].

Figure 1 shows, however, that a level budget does not necessarily mean a hollow program. There were several great accomplishments in the mid-70s to mid-80s when the space budget was declining, flat, or barely rising. The agency has become accustomed to claiming that all new initiatives depend on budget increases, but an increase is not likely to happen and history shows it is not necessary.

Further, if the NASA budget levels off at about $14 billion, that is a great deal of money. Any assertion that $14 billion spent every year will not fund significant space accomplishments is ridiculous and should not be treated seriously. If most of the money is being spent for "operations," i.e. for the continuing support of infrastructure, then we must examine the productivity of the operations before looking for a bigger budget. (Whether $14B is enough take us to Mars while maintaining a balanced program is a question to be answered when the nation becomes serious about the attempt.)

But beyond the obvious fact that space funding is adequate for now, what can be done constructively? Or, put another way, before program managers are given more money, how can they better manage what they have?

Clearly there is a need for the space program to change in fundamental ways. It must change and it will change either for better or for worse. New players in the private sector and in other agencies of government are struggling to take roles. If peace truly comes the DoD space actors may broaden their role. The space program as we know it will die or wither away if it does not at least accompany and accommodate change. Best would be for NASA to lead the way.

There is a need to change the culture or psychology of the whole spae program, that is NASA, contractors, Congress, and White House. There needs to be a new challenge: To do more for less, the challenge of a level budget. Cost needs to be made a major criterion. For example, science missions could be competed on the basis of the best science for a given amount of funding – and the discipline would have to be found to enforce such a spending cap. That is the beauty of a level budget – it becomes clear that overruns in one project directly hurt another.

There needs to be another, related, new challenge for programs to perform as promised. When the promises made during the "selling" of a program are not taken seriously during its subsequent execution, there is no reason for the program's promoters to take them seriously at any time: That is, the operating principle becomes to promise whatever you can get away with, which cheapens and tarnishes the whole program. Thus there needs to be a dose of honesty and accountability injected into the agency's culture.

Competition needs to be based on performance. For science missions performance criteria should be scientific. For boosters they should be cost, reliability, and ease of use. For applications criteria should be developed by the external user communities. For technology development programs the criteria should be set by the potential users. When these users are within the agency great attention must be paid to assuring that there is open debate about the real need for such developments; external oversight is indispensable to avoid "hobby shop" operations. For exploration programs NASA is the agent for the whole U. S. citizenry – NASA has our proxy – and so the criteria must be political in the best sense of the word. Apollo was clearly endorsed by the American people – particularly through their elected representatives. Such endorsement should be sought for the return to Mars.

And there needs to be more competition. Not the so-called "competition" that is carried out under the Federal Acquisition Regulations. That is a regularized and accepted way of conducting procurements but it has little connection to the economic concept of competition. For example, it has little to do with performance and everything to do with promises, and we have seen that the two have little to do with each other.

There needs to be competition based on scientific or practical criteria as discussed above with cost a major factor. We have had competition with the Soviets which was real, but it was for prestige and thus more concerned with image than with substance. It led to a kind of potlatch approach to space where the contest almost seems to have been based on who could spend the most money. We may now find serious substantive competition from our foreign allies in Europe and Japan. We could develop healthy performance-based competition between NASA centers to replace the present unhealthy competition for resources. (That is, they could compete on outputs rather than for inputs.)

In other words, there is a need to change the context and process inside NASA — which is largely within NASA's control. The papers in this volume have shown that program performance is largely driven by such context and process.

Where Do We Go From Here?

What can the nascent space policy research community do? There is a need to continue to accumulate research results. The space policy literature is thin and many of the problems[5] are virtually untouched. Only three will be mentioned here to give a flavor of what can be done.

Costs are more important in this era of unprecedented Federal deficits not yet under control. It is no longer reasonable to assume that costs will be absorbed or discounted in favor of technical performance. If a project is allowed to exceed its budget, there will be large opportunity costs as other projects are postponed, curtailed or cancelled; large political costs as political "chips" are expended to secure extra funds; or a reduction in program flexibility caused by mortgaging the future.

Traditionally the role of the political system has been to deliver the funds needed to carry out the vision of space exploration: Mission design was specified first and then costs were determined by the amount of money needed to achieve the desired design. The system (driven largely by Federal procurement regulations) is required to be sensitive to annual appropriation ceilings but not to overall project costs. That is, cost overruns are pushed into future years, delaying accomplishments and sometimes making them obsolete.

The design of a science mission, for example, is typically determined by asking users "what do you need?" This generates a wish-list which is then used to justify building a large, expensive system to do many of the things on the list. (This preserves broad support in the user community.) To make cost a more significant criterion one must ask instead, "if you had a dollar, how would you spend it?" This very different question allows, indeed forces, fundamental consideration of alternatives for meeting needs.

Policy research is called for: How can we put incentives into the procurement system to hold down costs and get the most of each dollar invested? The Administration and Congress have promoted "commercialization" of space activities in part to reduce costs. However, it may be difficult to achieve the anticipated benefits of "commercializing" critical activities which the government will not allow to fail. For example, if the government is determined to build a Space Station then it cannot reasonably allow *critical* elements to be built commercially, if the definition of "commercial" includes some real risk. What other approaches might work? If budgets are going to be level, what insights, tools, and incentives can we give NASA managers to enable them to work productively? At present they do not have a useful tool kit.

Performance also must become increasingly important, and the criteria of performance is changing. NASA has a good record with respect to the ultimate technical performance of its missions but there is a disturbing tendency to defer delivery of results. Missions are taking longer and longer between initiation and culmination. Galileo began in 1976 but data from Jupiter will not be available before 1995, and it now appears that much less data will be delivered than was promised. Station "began" in 1984 and the first element launch is planned for 1995 with complete assembly of Block I in 1998, and Block I is only a fraction of what was envisioned in 1984. The deferral of payoffs into the future discounts the present value of missions both politically and economically, which makes program support and funding more problematical.

In addition, "technical" criteria of success are becoming less important: It is not enough merely to fly hardware successfully; *de facto* standards will be set through competition. The fact that the Shuttle is a great technical success and copied by the Soviets is less important than whether users believe that it can take their payloads into space reliably and efficiently, compared to the alternatives. One user, the U.S. Air Force, has decided that the Shuttle is not its vehicle of choice for many payloads[6]. In the beginning it was enough merely to get into space. But reliability and efficiency become more important once the uses of space have been demonstrated.

In summary, our space program has emphasized technical challenge and technical performance – Oppenheimer's "sweetness" – over real and solid utility. The policy research problem is to devise ways to stimulate and challenge the people in the program to achieve both technical improvement and cost-effective utility. Because the latter is more difficult, it could be even more challenging and stimulating.

Large projects take longer to develop, which means they may become technically obsolete before they are flown. In this way old technologies get locked into flight projects as Faget has shown. This together with insensitivity to cost may make such large projects irrelevant or even counterproductive both to U. S. industrial competitiveness and to the education of the next generation of engineers and scientists.

Large scale goals do not necessarily require large-scale, centralized projects. Some large-scale goals can be realized efficiently by small evolutionary steps which take advantage of feedback to delete unpromising alternatives and to develop promising ones. In situations where we can neither predict nor control the future of a project nor the environment on which it depends – i.e. in typical situations – such an evolutionary approach is probably the best course.

McCray and Stern and Coleman have called for smaller science missions, countering earlier assertions[7] that the nature of space science makes larger missions inevitable. They also give in their chapters some fairly specific diagnoses of what is wrong and prescriptions for how to fix the situation. More recently NASA's space science advisory committee has recommended more emphasis on small missions[8]. Thus this research problem may have evolved into the need to monitor what happens. Or we may need to develop policy incentives to ensure that the recommendations are realized. Projects like Station might be kept small by empowering users – for example by expanding the opportunity for them to advocate smaller projects that are quicker to fruition. But how does one so empower users?

Some projects, such as a manned mission to Mars, are unavoidably large. A research question is how to modularize them so they can be accomplished effectively even if problems arise. Initially, modules may cost a little more in the short-run, but in the long-run might avoid the domino effect of costly failures rippling through a tightly integrated system. How much modularity is enough, and how do we achieve it?

These are the old research problems, framed in the old terms; cost, performance, project size. But as we have seen the root causes for the manifestations – i.e. cost overruns, schedule delays, giantism – lie in the culture and incentives of the program. These are likely to be effectively addressed by research supplemented with the ideas and methods of anthropology, sociology, psychology, and management. Future research might profitably tap these disciplines. For example, how can individual incentives be structured to achieve desired results – e.g. smaller science missions – and also fit into a civil service personnel system?

There is a need to understand the culture of the space program in terms of theoretical concepts or models that lead to prescriptions for constructive change. We need to know how to define incentives that will be effective in creating a better program, one that is aiming at the right goals and accomplishing what it promises with the resources it gets.

Probably the incentives for individuals should not be principally economic. That is, the incentives that will pull out the extra creative effort for a space exploration will not be a few more dollars. Especially in government this is unlikely to work. Mainly economic incentives are likely to stimulate the wrong responses. We want to stimulate creativity not cupidity. This will involve creating tough technical challenges, but there needs to be a credible chance of success.

Finally there remains a continuing need for vision and leadership. These qualities must address not only the outward goals of the space program but also its interior execution. Today there is great need for change in the culture of the program, and this calls for as much vision and leadership as any space mission. Until program leaders want such change we are likely to have only the change of decay. With leadership toward the vision of a rejuvenated space program and with appreciation of the need for policy research and the application of the results of that research, the US can certainly conduct a vigorous civilian space program.

Notes

1. Caldwell Johnson as quoted in H. S. F. Cooper, Jr., "Annuals of Space, We Don't Have to Prove Ourselves," *New Yorker*, September 2, 1991, p. 50.

2. The House Appropriations Committee reported out a bill that zeroed funding for Station in fiscal year 1992. Normally such actions are adopted and it was expected that the Station would be killed. However, a floor amendment restored the full funding for the Station and the Senate concurred. See *Congressional Record*, June 6, 1991, p. H4021.

3. *Report of the Advisory Committee on the Future of the U.S. Space Program*, N. Augustine, chair, U.S.G.P.O., Washington, DC, 1990, p. 11.

4. This figure was provided by Dr. William S. Smith of the House Committee on Science, Space, and Technology.

5. R. Byerly and R. D. Brunner "Future Directions in Space Policy Research", in *Space Policy Reconsidered*, R. Byerly, Editor, Westview Press, Boulder, 1989. There is a discussion of eleven overlapping research areas on pp 178-187. The discussion in this chapter is adapted from material there.

6. This is reflected in their decisions to develop, purchase, and use the Titan IV Complementary Expendable Launch Vehicle (which is approximately equivalent, i.e. complementary, to the Shuttle) and the Delta II and Atlas-Centaur II. See B. Davis, "With its Titan IV, Air force at Last Takes Helm of Space Program, Putting NASA in the Backseat" *Wall Street Journal*, November 29, 1988, p. A20.

7. NASA Advisory Council, Space and Earth Science Advisory Committee, *The Crisis in Space and Earth Science*, Washington, D.C.: NASA, November, 1986.

8. B. Moore III, Chair, Space Science and Applications Advisory Committee, letter dated August 27, 1991 to L. Fisk, Associate Administrator for Space Science and Applications, NASA HQ, Washington, DC. this letter was the report of a five-day retreat to plan space science missions.

INDEX

ABM. *See* Anti-ballistic missile treaty
Accountability, 95, 96, 98, 102, 104, 142
Acquisition regulations, 123
ACRV. *See* Assured Crew Return Vehicle
Active Magnetospheric Particle Tracer Experiment (AMPTE), 147
Advanced Communication Technology Satellite program, 14, 168
Advanced Launch System (ALS), 31, 32
Advanced X-ray Astrophysics Facility (AXAF), 30, 101, 102–103, 105, 145, 154, 160, 179, 181(n1)
Aerospace industry, 14, 45, 58, 132
  incentives for entry, 107, 115–116
  participation by, 112–113
  stock performance, 108–109
  See also Contractors
Ahearne, John F., 45, 48–49(n26)
Air Force, 28, 32, 117, 118, 271
Alaska SAR Facility, 198(n10)
ALS. *See* Advanced Launch System
American Satellite, 23
American Telephone & Telegraph, 23
Ames Laboratory, 118
AMPTE. *See* Active Magnetospheric Particle Tracer Experiment
Amroc, 132
Anik system, 23
Anti-ballistic missile (ABM) treaty, 19, 32
Apollo paradigm, 2–3, 5, 265–266
Apollo program, 38, 40, 170, 180
  cost of, 39, 239(table), 245(n36)
  goal of, 239–240
  moon mission, 32, 93, 120–121, 124
  myths related to, 1

1967 fire, 44, 123
  schedule, 46, 119, 122
Apollo-Soyuz Test Project, 38, 121, 239(table)
Arabsat, 24
Ariane rocket, 32
Army, 117, 118
Assured Crew Return Vehicle (ACRV), 110
ASTRO-D, 160
Astronauts. *See* Manned spaceflight
Astronomy, 30
Astrophysics, 173(fig.), 180
Atlantis, 238(table), 244(n29)
Atlas-Centaur launch vehicle, 120, 273(n6)
Augustine, N., 3, 4, 5–6, 7
Augustine Report, 53, 70, 75(n20), 139
Australia, 24, 25
"Award fee" system, 107
AXAF. *See* Advanced X-ray Astrophysics Facility

Balanced Budget and Emergency Deficit Control Act. *See* Gramm-Rudman-Hollings
Bay of Pigs invasion, 36
Beggs, James, 94, 95
Bell companies, 23
Benchmarks, 62, 71–72, 105, 113, 154
Bid and proposal (B/P) funds, 111, 115, 116
Biogeochemical cycles, 187, 189(table)
Biosat, 133
Boeing, 45, 132
B/P. *See* Bid and proposal funds
Brazil, 24
Brunner, R., 5

275

Budget
 battles, 40, 44, 45, 53–55, 60, 69–70, 85–91, 176–178, 266–267
 history, 175, 176(fig.)
 multi-year appropriations, 164
 planning, 151–154, 164
 resource allocation, 159–160, 164
Budget Committee, 83
Bush, George, 9(n5), 14, 32, 44–45, 79, 96, 114, 183, 188, 190

Canada
 communications satellite system, 23
 space science program, 182(n7)
Carroll, Lewis, 139
Carson, Kit, 253
Carter, Jimmy, 17
Cassini Saturn mission, 29, 101–102, 145, 154, 179
CBO. *See* Congressional Budget Office
CCIR, 25
Centaur launch vehicles, 30, 33(n2)
Central Pacific railroad, 252
Centrifuge uranium enrichment plant, 70
Challenger accident, 29, 93, 114, 144, 171, 180, 197, 233, 235(&fig.), 238(table), 259, 266
 budgetary effects of, 40, 44, 152, 155
 causes of, 213, 222(n55)
 policy effects of, 2, 5–6, 123, 142, 143
 replacement costs, 85, 231, 244(nn 20, 24, 26)
"Changing University Role in Space Research, A," 157–158
China, 24, 32
CIESIN, 198(n11)
Climate change, 183, 184, 187, 188, 189(table), 190. *See also* Global Change Research Program
Clinch River Breeder Reactor, 70
COBE. *See* Cosmic Background Explorer
CoF. *See* Shuttle program, Construction of Facilities

Cold War, 14, 15, 35, 39, 40, 41, 46, 47. *See also* Soviet Union
Columbia, 235(&fig.), 238(table), 245(n30), 251
Columbus, Christopher, 255, 260
Comet Rendezvous-Asteroid Flyby/Cassini missions, 29, 101–102, 145, 154, 179
Commercialization, 129, 132, 133–135, 264–265, 268, 270
Communications Satellite Act of 1962, 20–21, 22
Communications satellites, 19–20, 33, 155
 direct broadcast (DBS), 25
 domestic service, 22–24
 international, 20–21
 markets, 13, 14, 24
 military, 16
 mobile service, 21–22
Competition, 268–269
Comsat Corporation, 25, 27
Congress, 6, 10(n13), 36–37, 44–45, 159, 164, 222(n56), 252, 266, 268
 appropriations committees, 55, 81, 83, 84, 87
 budget process, 54–55, 79–81, 83–87, 90–91(n6), 91(n7)
 funding and, 53–54, 56, 58, 60, 79, 154
 large-scale bias of, 147, 148
 1980 space science crisis and, 141, 145
 objectives of, 66
 oversight by, 98, 100–101, 102–103, 113–114, 116
 Space Station and, 204–206, 210–211, 216–217
Congressional Budget Office (CBO), 77
Conspicuous consumption, 42
Constituencies, 97–98
Contractors, 60, 61, 74(n9), 107, 266, 268
 defense, 45–46
 monopsony, 108–109
 procurement process and, 109–112
 support service, 123

See also Aerospace industry
Coronado, Francisco Vasquez de, 260
Cortes, Hernando, 260
Cosmic Background Explorer (COBE), 67, 147, 155–156
Cost, 269–270, 271
 allowances, 112
 containment, 162–163
 failure and, 152
 modularization and, 271–272
 overruns, 74(n13), 95
 projections, 96–97, 148, 152, 154
 of single-mission programs, 181
 of small missions, 129–130, 135
 "sunk," 70–71
 unconstrained, 148
 upward spiral of, 127–128, 130, 164
Cost-benefit analysis, 67, 68(table), 69–72, 73–74(n7), 146–147
CRAF. *See* Comet Rendezvous-Asteroid Flyby/Cassini missions
"Crisis in Space and Earth Science, The." *See* Space and Earth Science Advisory Committee report
Cuba, 36, 39
Custer, George Armstrong, 253
Czechoslovakia, 39

DAACs. *See* Distributed Active Archive Centers
DARPA, 130, 132
Data management, 186–187, 190, 192–195
DBS. *See* Communications satellites, direct broadcast
Defense contractors, 45–46
Defense spending, 14, 80, 82(table), 86(table), 89
Delta launch vehicles, 30, 33(n2), 130, 273(n6)
Department of Defense, 14, 15, 17, 185, 228, 268. *See also* Military Program
Department of Housing and Urban Development, 84, 85, 87
Department of Transportation, 63

Department of Veterans Affairs, 84, 85, 87
Discovery, 235(&fig.), 238(table), 244(n29)
Discretionary programs, 80, 81, 82(table), 86(table), 88(table), 89
Distributed Active Archive Centers (DAACs), 194–195, 198(nn 10, 11)
Documentation requirements, 151
DSCS spacecraft, 16
Dust Bowl, 256
Dyson, Freeman, 121–122

Earth Observing System (EOS), 25–28, 69, 71, 72, 102, 103, 117, 131, 149, 180, 218, 264
 budget, 79, 152, 188, 189(table)
 Data and Information System (EOSDIS), 185, 186–187, 190, 192–195
 Frieman report on, 4, 5
 planning, 196–197
 role of, 184, 189–191, 193–194, 195–197
 scale of, 191
 science program, 187–188
 space-based system, 185
Earth probes, 145, 184, 185, 186, 188, 194
Earth sciences, 173(fig.), 180
Earth System Science, 187, 189(table)
East Germany, 39
Echo series, 168
Ecosystems, 187, 189(table)
Edwards Air Force Base, 234
Egypt, 42
Einstein satellite, 160
Eisenhower, Dwight, 35–36
ELVs. *See* Expendable launch vehicles
Endeavor, 231, 244(n26)
Emerging democracies, 17
Energia launch vehicle, 31
England, 25
Entitlements, 80, 82(table), 86(table), 89
Environment
 atmospheric oscillations, 185, 187

global changes, 183, 184, 187, 188, 189(table), 190
human interactions, 189(table), 190, 193
oceanic circulations, 187
Environmental Protection Agency, 85
EOS. *See* Earth Observing System
Eosat, 27
EOSDIS. *See* Earth Observing System, Data and Information System
EROS Data Center, 198(n10)
ESA. *See* European Space Agency
Europe
competition from, 269
EOS and, 185
space programs, 14–15, 33
unification of, 46
U.S. forces in, 16
European Space Agency (ESA), 27, 41, 131
Eutelsat, 24
EUVE. *See* Extreme Ultraviolet Explorer
Expendable launch vehicles (ELVs), 6, 10(n14), 33(n2), 130, 132, 143, 144, 155, 156
Explorer Program, 103, 145, 162, 169
Exploration, 29, 264–265. *See also* Human Exploration Initiative
Extreme Ultraviolet Explorer (EUVE), 156
Exxon Valdez, 256

FAA. *See* Federal Aviation Administration
Far Ultraviolet Spectrometer Explorer (FUSE), 156
Federal Acquisition Regulations, 268
Federal Aviation Administration (FAA), 17
Federal Communications Commission, 22, 25
Federal deficit, 8, 15
Federal Express, 23
Feynman, Richard, 259
Fiber optic cables, 13, 20, 34(n5)
Fisk, Lennard, 140, 144

Fleetsatcom, 16
Fletcher, James C., 201, 223
Ford (motor company), 23
France, 39
comsat system, 24
earth observation by, 26
Freedom. *See* Space Station
Fremont, John C., 260
Frey, Louis, 176
Frieman, E., 4, 5
Frontier model, 250, 252–257, 258
Fulbright, J. W., 1
Funding
concentration of, 142–143, 144–149
multi-year, 115
planning and, 152–154
See also Budget; Congress
FUSE. *See* Far Ultraviolet Spectrometer Explorer

Gagarin, Yuri, 36, 37
Galbraith, J. K., 53
Galileo mission, 29, 140, 145, 150, 151, 179, 270
Gamma Ray Observatory (GRO), 30, 140, 167
GAS. *See* Getaway special
G booster, 37
Gemini Program, 119, 120, 121, 122, 239(table)
General Electric, 23, 45
General Revenue Sharing, 81
General Telephone, 23
Geology, 187
Geophysics, 187
Germany, 17, 46. *See also* East Germany
Getaway special (GAS), 133
GFE. *See* Government-funded equipment
Glenn, John, 36, 120
Glennan, T. Keith, 118
Global Change Research Program (USGCRP), 104, 183–184, 187, 188, 189(table), 189–191, 193–194, 195, 197
Global positioning system (GPS), 17

Glonass, 17
Go-as-you-pay principle, 5, 6
GOCOs. *See* Government-owned, contractor-operated laboratories
Goddard Space Flight Center, 194, 198(n10)
Golden Spike, 251, 252
Gore, Albert, 221(n48)
Government-funded equipment (GFE), 133–134
Government-owned, contractor-operated laboratories (GOCOs), 64
GPS. *See* Global positioning system
Gramm-Rudman-Hollings (GRH), 77, 83, 85, 87, 89, 90(n5), 263–264
Green, Bill, 217, 221(n50)
Greenhouse effect, 183. *See also* Climate change
GRH. *See* Gramm-Rudman-Hollings
GRO. *See* Gamma Ray Observatory
Guastaferro, Angelo, 63

Halley's comet, 149
Hearth, Don, 110
Hearth Study, 110–111
HELIOS program, 168
High Performance Computing and Communications Initiative, 104–105
Hinners, Noel, 175–176
HST. *See* Hubble Space Telescope
Hubble Space Telescope (HST), 15, 30, 44, 93, 101, 102, 140, 145–146, 147, 150, 152, 155, 167, 179, 197, 221(n38), 222(n52)
Hughes Communications, 23
Human Exploration Initiative, 3, 9(n5), 14, 29, 72, 93, 96, 114, 117, 118, 122, 218
Hungary, 39
Hydrologic cycle, 187, 189(table)
ICBMs, 37
Incentives, 95–104, 263, 271, 272
Independent research and development (IRAD), 111, 115, 116
India, 24, 39

Indonesia, 23
INF. *See* Intermediate-range Nuclear Forces treaty
Information
 budget process and, 61
 public, 55
 See also Data management
Infrared Space Observatory (ISO), 149
Infrastructure, 97
Inmarsat, 21, 27, 33
Institutional reform, 63–64
Intelligence. *See* Military Program, surveillance
Intelsat, 17, 20–21, 26, 33
Interagency research efforts, 104–105
Interest, 80, 82(table), 86(table)
Intermediate-range Nuclear Forces (INF) treaty, 18, 19
International Civil Aviation Organization, 17
International collaboration, 69
International Ultraviolet Explorer (IUE), 155
IRAD. *See* Independent research and development
Iridium system, 22
ISAS. *See* Japan, Institute of Space and Astronautical Sciences
ISO. *See* Infrared Space Observatory
Israel, 33(n4)
IUE. *See* International Ultraviolet Explorer

Japan, 8, 46
 ASTRO missions, 160
 competition from, 269
 comsat system, 24
 DBS spacecraft, 34(n7)
 EOS and, 185
 Institute of Space and Astronautical Sciences (ISAS), 160–161
 launch vehicles, 32
 space program, 15, 33, 41, 178, 182(n7)
Jet Propulsion Laboratory (JPL), 64, 133, 198(n10)
Job creation, 66

Johnson, Caldwell, 35, 263, 266
Johnson, Lyndon, 17, 35, 36, 38, 39
Johnson Space Center, 124, 228
JPL. *See* Jet Propulsion Laboratory

Keck telescope, 30
Kennedy, John F., 36–37, 38, 39, 95, 121
Kennedy Space Center, 39, 160, 228, 234
Killian, James R., 36
King, Clarence, 260
Knowledge, 71–72. *See also* Data management; Information
Korea, 16

Landsat, 26–27, 28
Landsat Remote Sensing Commercialization Act of 1984, 27
Langley exercise, 201, 202, 212–213
Langley Research Center, 118, 198(n10)
Lanzerotti, L. J., 157–158
Large-scale missions, 147, 148, 164, 169–182, 271–272
Lasswell, H. D., 223
Launch rates, 144, 145, 154, 159–160, 170, 172(fig.), 173(fig.), 180
Launch vehicles, 30–33, 134
  evolution of, 126–127
  expendable, 6, 10(n14), 33(n2), 130, 132, 143, 144, 155, 156
  low-cost, 132
  markets, 13–14, 15, 16
  for scientific missions, 143
  selection of, 156
  success rates, 239(&table)
Lenoir, William B., 201–202
Lewis, Meriwether, 260
Lewis Laboratory, 118
Lindbergh, Charles, 74
Livy, 223
Lockheed, 45, 132
Lodestar launch vehicle, 132
Long March launch vehicle, 32
Lunar GAS satellite, 133
Luxembourg, 25

Macauley, M. K., 229, 243(nn 8, 9)
McCormack, John, 37
McCray, Richard, 179
McDonnell Aircraft Corporation, 118, 120
McDonnell Douglas, 45
McDougall, Walter A., 35
Magellan mission, 29
Management authority, 123, 124
Manned Orbiting Laboratory (MOL), 33(n1), 212–213
Manned spaceflight, 2, 29, 97, 264
  program accountability, 98–101
  program costs, 239(table)
  prospects for, 45–47
  R&D, 45
  reasons for, 35, 41, 46, 47
  space science linkage to, 154–156, 164
Marshall Space Flight Center, 124, 198(n10), 228
Mars mission, 29, 39, 41, 44, 45, 63, 64, 163, 264, 266, 271
Martin Marietta, 23, 45
Math and Science Education Initiative, 105
Means vs. ends, 70
Media, 55
Medium launch vehicles (MLVs), 130
Mercury-Friendship, 7, 36
Mercury Program, 118, 119, 120–121, 122, 239(table)
Merit review process, 102
Mexico, 24
MfPE. *See* Mission from Planet Earch
Microsat, 132
Mikulski, Barbara, 199
Military Program, 13, 15–16, 33(&n1), 45–46
  communications, 16
  defense, 19
  industry and, 112
  missile warning, 18–19
  navigation, 17
  surveillance, 17–18
  See also Department of Defense
Milstar system, 16

MIR space station, 43–44
Mission from Planet Earth (MfPE), 3, 4, 6
Missions, 1, 2
  complexity of, 79, 181
  development phase, 142
  discipline distribution, 169–171, 173(fig.), 174(fig.), 177, 179, 180
  funding, 185
  large-scale bias, 147–149, 164, 169–182, 271–272
  lead times, 145, 147, 157, 158, 175, 179, 270
  level budgets for, 268
  requirements, 97
  scientific, 29–30, 159–160. *See also* Space science program
  selection of, 126
  small, 125–136, 142, 145, 149
  stretch-outs of, 152–153
Mission to Planet Earth (MtPE), 3, 4, 67, 114, 184, 188, 192, 194
MLPs. *See* Mobile Launch Platforms
MLVs. *See* Medium launch vehicles
Mobile Launch Platforms (MLPs), 234
Modules, 271–272
MOL. *See* Manned orbiting laboratory
Mongolis, 39
Monopsony, 108
Moon mission, 1, 2, 3, 32, 36–38, 41, 44, 93, 95, 120–121, 124, 169, 170
Motorola, 22
Moynihan, Daniel Patrick, 221(n50)
MtPE. *See* Mission to Planet Earth
Multi-center programs, 124
Myers, Dale, 212, 214, 216

NACA. *See* National Advisory Committee for Aeronautics
NAR. *See* Non-Advocate Review
NASA. *See* National Aeronautics and Space Administration
National Advisory Committee for Aeronautics (NACA), 117
National Aeronautics and Space Administration (NASA), 14, 35
  accomplishments, 267(&fig.)
  benefit distribution, 73–74(n7)
  budget battles, 40, 44, 45, 53–55, 60, 69–70, 85–91, 176–178, 266–267
  budget history, 175, 176(fig.)
  core program, 77–79, 90(n2)
  creation of, 117
  field centers, 109–110
  institutional failure of, 258–259
  large-project bias, 271–272
  oversight of, 268. *See also* Congress; President
  priorities of, 3–4
  reorganization of, 64
  Space Science and Applications Advisory Committee, 4
  See also Office of Space Science and Applications
National Center for Atmospheric Research, 198(n10)
National Institutes of Health (NIH), 81
National Oceanic and Atmospheric Administration (NOAA), 26, 28, 185, 193, 195
National Research Council (NRC), 103
National Science Foundation (NSF), 54, 64, 81, 85, 197
National Snow and Ice Data Center, 198(n10)
National Space Act, 117
National Space Council (NSpC), 7, 8, 14, 114, 266
National Technical Means (NTM), 17, 18, 33(n4)
Naugle, John, 175
Navy, 117, 118
Nelson, Bill, 205
Newell, Homer E., 175
NIH. *See* National Institutes of Health
"90-day study." *See* Human Exploration Initiative
Nixon, Richard, 17, 40, 45, 96
Non-Advocate Review (NAR), 110, 115
North Vietnam, 39
NRC. *See* National Research Council; Nuclear Regulatory Commission
NSF. *See* National Science Foundation
NSpC. *See* National Space Council

NTM. *See* National Technical Means
Nuclear Regulatory Commission (NRC), 90–91(n6)
Nuclear Test Ban Treaties, 18

OAO. *See* Orbiting Astronomical Observatory
Observatories, 30, 150, 167
Odom, James B., 201, 210, 216
Office of Management and Budget (OMB), 141, 148, 154, 159, 217
Office of Space Science and Applications (OSSA), 140
   budget, 175, 176(fig.)
   cost projections by, 152
   large-mission bias, 147–149, 169–182
   Planetary Division, 163
   Strategic Plan, 141, 143–144, 151–152, 156, 159
OMB. *See* Office of Management and Budget
Open Skies Policy, 22–23
OPFs. *See* Orbiter Processing Facilities
Orbital Science Corporation, 130, 132
Orbiter Processing Facilities (OPFs), 233
Orbiters. *See* Shuttle program
Orbiting Astronomical Observatory (OAO), 150, 167
Orbiting Solar Observatories, 150
OSSA. *See* Office of Space Science and Applications
"Other Side of the Moon, The," 259
Ozone layer, 183

Paine Commission Report, 254–257
"Pay as you go," 89
Peacekeeper missiles, 130
Pegasus launch vehicle, 130
Perestroika, 44
Performance, 268, 269, 270–272
Pioneering the Space Frontier, 254–257
Pioneer programs, 175
Planetary Observer, 145
Planetary science, 29, 174(fig.), 180
Planning, 96, 126

budgetary, 151–154, 164
long-term, 104–105
Poland, 39
Policy
   change in, 266–269
   contexts, 264
   debate over, 6–8, 64–72
   research, 269, 270
Post-Challenger paradigm, 2–5
Potlach, 43, 269
President, 164, 266, 268
   large-scale bias of, 147, 148
   oversight by, 98, 100–101, 102–103
Principal-agent problem, 55–60, 73(n3)
   correction of, 62–64, 69–72
   resultant behaviors, 60–62
Principles of Project Management (Beggs), 94–95
Procurement, 74(n9), 109–112, 123
   commercialization of, 129, 132, 133–135, 264–265, 268, 270
   phased, 109–112
Project approval, 95
Program advocacy, 109–111, 112–113
Program evaluation, 105
Program pacing, 121–124
Proton launch vehicle, 31, 32
Public debate, 6–8, 64–72
Public opinion, 56, 93–94, 123, 268
Pyramids, 41–42

QA&R. *See* Quality assurance and reliability programs
Quality assurance and reliability (QA&R) programs, 127–128, 129–130
Quayle, Dan, 7, 14

R&PM. *See* Shuttle program, research and program management
Reagan, Ronald, 19, 44, 79, 80, 85, 96, 251
Remote Land Sensing Commercialization Act of 1984, 14
Report of the Advisory Committee on the Future of the U.S. Space Program. *See* Augustine Report

Research and Development, 107, 114
    federal spending for, 54, 56, 57(table), 58, 59(table)
    independent (IRAD), 111, 115, 116
    innovative management of, 63–64
    "successful," 61–62
    sunk costs, 70–71
    university-based, 125–126, 128, 131, 134–135, 136, 157–158
Resource distribution, 98, 142–143, 144–149
Risk
    assessment, 161
    avoidance, 97, 149–151, 164
    of large missions, 145–146
    of single-mission programs, 175–176, 178, 179
Roe, Robert, 204–205, 219(n13)
Roentgen Satellite (ROSAT), 168
Rockwell, 45
Romania, 39
ROSAT. *See* Roentgen Satellite
Rosendhal, J. D., 157–158

Sanford, Leland, 251
Satellite Business Systems, 23
Satellites
    communication, 13, 14, 16, 19–25, 33, 155
    cost of, 160
Saturn launch vehicle, 32, 40, 48(n17), 93, 98, 124
Savings & Loan crisis, 222(n56)
Schedule slippage, 95
Schumer, Charles, 221(n50)
Science
    community, 58, 72
    public opinion of, 74(n8)
    spending for, 59(table)
    See also Earth sciences; Space science program
Scout missiles, 30
SDI. *See* Strategic Defense Initiative
SE&I. *See* System Engineering and Integration
Selective inattention, 220(n22)

SESAC. *See* Space and Earth Science Advisory Committee report
SFCDC. *See* Shuttle program, space flight, control and data communications
Shakespeare, William, 183
Shepard, Alan, 36, 43, 120
Shuttle C, 31
Shuttle program, 3, 29, 39, 44, 45, 46, 49(n28), 67, 93, 96, 98, 117, 119, 124, 270–271
    capability, 225, 236–240, 241(table), 242
    construction of facilities (CoF), 228, 229
    cost, 156, 224, 225, 226(table), 227–233, 236, 239(table), 240, 243(nn 9, 12), 244(n22)
    cost-benefit analysis, 75(n24)
    dependence on, 2, 4–5, 10(n14), 102, 141, 144, 154–155, 156, 223
    flight rate, 121, 122, 225, 226, 233–235, 236, 237(fig.), 238(&fig.), 240, 245(n35)
    launch problems, 15
    low orbit problems and, 155, 156
    payloads, 240
    performance shortfall, 99–100, 224, 226–227, 236, 237(fig.), 241(&table)
    R&D, 227, 229
    reimbursements, 228–229
    reliability, 231–232
    research and program management (R&PM), 228, 229
    schedule factors, 233–234
    space flight, control and data communications (SFCDC), 227–228, 229
    space station and, 207, 208, 209
    success measurement, 61
Shuttle Transfer Aircraft, 234
Sigma Xi, 58
"Significance of the Frontier in America, The," 253
SIRTF. *See* Space Infrared Telescope Facility

602(b) allocation, 84, 87
Skylab, 48(n17), 93, 98, 121, 168, 171, 217, 239(table)
Small Explorers (SMEX) missions, 130, 132, 140, 143–144, 145, 159
Social spending, 59(table)
Solar and space physics, 174(fig.), 180
Southern Pacific Communication, 23
Southern Pacific Railroad, 252
Soviet Union, 14, 15, 16, 17, 18, 27
    collapse of, 1–2, 8, 19
    earth observation by, 26
    launch vehicles, 31–32
    manned program, 43–44
    military space program, 17, 18
    space race with, 1–2, 3, 35–39, 40–41, 43, 44, 117, 269
    space science program, 178
Soyuzkarta, 27
Space access, 125–136
Space and Earth Science Advisory Committee (SESAC) report, 141, 144, 156
Space Corporation, 132
Space Council. *See* National Space Council
Spacecraft complexity, 131
Space defense, 19
Space Exploration Initiative. *See* Human Exploration Initiative
Space Frontier Agency, 63, 64
Space Infrared Telescope Facility (SIRTF), 30, 110, 149, 181(n1)
Spacelab, 15, 143, 144, 155, 159, 168, 171
Space manufacturing, 44
Space policy. *See* Policy
Space race, 35–41, 43, 117
Space science program, 3, 4, 6, 10(n13), 70, 75(n20), 114, 139–142
    crisis in, 141, 145
    discipline vitality of, 147
    diversity of, 143, 177, 178
    documentation requirements, 151
    expertise for, 157–159
    management responsibility in, 161–164
    mission budgets, 142
    mission scale and, 175–180, 271
    mission selection, 148
    threats to, 144–159
Space Station, 2, 3, 4, 6, 10(n16), 29, 31, 39, 44, 46, 58, 93, 96, 97, 98, 117, 119, 124, 133, 271
    accountability, 100, 101
    budget, 201, 203–204, 206–207, 209–210
    Congress and, 204–206, 210–211, 216–217
    contracts, 111–112
    controversy around, 199–200
    disruptions, 207–209
    element decoupling, 215–217, 219(n5)
    funding, 87, 219(n15), 220(n27)
    lead time, 270
    plasma problem, 208
    program pace, 121, 122
    program redesign, 104, 203–205, 213–217, 218
    resilience of, 205–209, 215–217
    Shuttle dependency of, 208, 209
    stability, 200–209, 210–211, 212, 213, 215–217, 218, 219(n6)
    termination of, 209–213, 214, 218, 266, 273(n2)
Space Task Group, 118, 120
"Spillovers," 66, 67. *See also* Commercialization
SPOT system, 26, 27
Sputnik, 35, 37, 117
START. *See* Strategic Arms Reduction Treaty
Stern, S. Alan, 179
Stofan, Andrew J., 200–201
Strategic Arms Reduction Treaty (START), 16, 18
Strategic Defense Initiative (SDI), 15, 19, 31, 32, 33
STS-5, 121
Stubbing, Richard A., 45, 48–49(n26)
Subcontractors, 161
Suborbital programs, 103
"Success," 61–62

criteria, 270–271
measurement of, 71
orientation toward, 150–151, 259
of unmanned missions, 176–177, 177(fig.), 181
Sunk costs, 70–71
Surveyor missions, 150, 169
Sweden, 182(n7)
Synthetic Fuels Corporation, 70, 81
Syria, 39
System Engineering and Integration (SE&I), 122

Tanaka, Yasuo, 160–161
Taurus launch vehicle, 130, 132
Taxes, 56, 57(table), 72
TDRS System, 6, 155
Technology
  commercial applications, 264–265, 268, 270
  critical, 153
  evolution of, 126–127, 132, 133
  expertise shortage, 164
Telecom spacecraft, 24
Telecommunications industry, 13
Telson, M., 54–55
Tiros weather satellites, 168
Titan launch vehicles, 30, 33(n2), 273(n6)
Titov, Gherman, 36
Toman, M. A., 229, 243(nn 8, 9)
TOPEX oceanography mission, 140, 145
Transportation, 2, 5, 56, 64, 154, 223
Traxler, Bob, 206, 216
Truly, Richard H., 201, 203–204, 212
TRW, 45
Turner, Frederick Jackson, 253, 255
Twain, Mark, 93

UARS. *See* Upper Atmosphere Research Satellite
UHF Follow-on, 16
UHF service, 16, 22
Ulysses Solar-Polar spacecraft, 15, 29

Union Pacific Railroad, 251–252
United States Geological Survey (USGS), 193, 195
Universities, 125–126, 128–129, 131, 134–135, 136, 157–158
University of Colorado, 133
University of Wisconsin Space Science and Engineering Center, 198(n11)
Unmanned missions
  budget for, 13
  program accountability, 101–103
  See also Space science program
Upper Atmosphere Research Satellite (UARS), 140
USGCRP. *See* Global Change Research Program
USGS. *See* United States Geological Survey
Utility spacecraft, 134

VAB. *See* Vehicle Assembly Building
VA-HUD appropriations, 84, 85, 87
Vanguard program, 175
Veblen, Thorstein, 42–43
Vehicle Assembly Building (VAB), 233
Venus Radar mission, 163
Veterans Administration. *See* VA-HUD appropriations
Vietnam war, 39
Viking lander mission, 140, 176
Vision, 96, 100–101, 272
Volkmer, Harold, 203
Voyager mission, 29, 93

Warsaw Pact, 14, 17
Weather service, 25–26, 27–28
Welch, J., 4–5
Western Union, 23
Wiesner, Jerome, 36

X-ray astronomy, 160–161
X-ray Timing Explorer (XTE), 156
XTE. *See* X-ray Timing Explorer

Yeats, William B., 199
York, H., 13